A GUIDE TO CLIMBING AYERS ROCK

Marc Hendrickx

Connor Court Publishing

Published in 2018 by Connor Court Publishing Pty Ltd.

Copyright © Marc Hendrickx

Connor Court Publishing Pty Ltd
PO Box 7257
Redland Bay QLD 4165
sales@connorcourt.com
www.connorcourtpublishing.com.au

ISBN: 9281925826098

Cover Design by Maria Giordano

Cover illustration: Ayers Rock – Kuniya walk (rock climbing) from Wikipedia Commons

Printed in Australia.

"The moment you say that any idea system is sacred, whether it's a religious belief system or a secular ideology, the moment you declare a set of ideas to be immune from criticism, satire, derision, or contempt, freedom of thought becomes impossible."

Salman Rushdie

CONTENTS

DEDICATION

This book is dedicated to the seven million brothers and sisters of the Rock who have climbed it, especially former head Ranger Derek Roff MBE who successfully managed the Park for 18 years between 1968 and 1985 during a time of dramatic political and social upheaval; and Anangu man, Tiger Tjalkalyirri, who was one of the first climbing guides and entertained tourists into his old age. Their common sense, good humour, patience and understanding have been missing from the Red Centre for far too long. There should be a bronze statue of both of them at the base of the Climb.

ACKNOWLEDGEMENTS

Everyone I spoke with who had climbed the Rock seemed re-energised by recounting their experiences and most found it difficult to reconcile the wonder of their adventure with the sanctimony and misery of the new rules at Uluru. In times I was struggling to find time to collect facts and figures, or having to respond to or ignore outrageous irrational taunts from people who are unable to separate the simple act of enjoying the natural world from the politics of race, I would recall their joyous accounts and be invigorated. Thank you all!

I would like to thank staff at the Northern Territory Archives Service, National Archives of Australia, National Library of Australia, Macquarie University Library, NSW State Library, Australia National Museum and the XNatmappers for their assistance in obtaining documents and providing permission to re-use photographs and text excerpts. Thanks also to the Lutheran Archives for permission to use frames from Lou Borgelt's remarkable 1946 movie, and his summit photo. Thanks to Edna Bradley (Saunders) for permission to use a photo of the Petticoat Safari and for discussing her experiences at the Rock with me. A special note of thanks to Steven Baldwin at UKNP for doing his best to answer questions in a difficult administrative environment and for returning "The Last Logbook". And thanks to Bobby Roff for filling in details about times when the Park was a more enjoyable place to visit. Thanks also to family and friends for their patience with my obsession to get the truth out.

No thanks to the Canberra Bureaucrats (you know who you are) who provided the most superficial, bland and meaningless responses to my inquiries. The words "public" and "service" are not compatible with your tenure. No thanks to the Australian Academy of Science who banned me from their social media accounts after I dared challenge their politics. Perhaps they should be renamed the Academy for Political Science? No thanks to members of the Uluru Park Board who refused to respond to my letters about the climb, and delegated their job to under-resourced Park staff.

To the politicians who allowed the plan to ban the Climb to pass through the parliament, despite voicing strong support for it, and having the backing of the Australian public, you failed to stand up for what you said you believed in. You have let our Western values be trampled into the dust, values that have provided permanent settlement in the dry desert heart of our Nation, allowing millions to enjoy its beauty. No thanks for you or the current crop of ministers who failed to provide direct responses to my inquiries abrogating responsibility to their hollow-headed functionaries.

FOREWORD

One of Australia's national icons and tourist attractions is Ayers Rock. Visitors to Australia from abroad want to see Ayers Rock and the Olgas, the Great Barrier Reef and maybe the Great Ocean Road, the Blue Mountains, the Sydney Opera House and various native animals in the wild and in zoos. Visitors try to pack this into a fortnight. Most visitors to Ayers Rock want to climb it, most visitors stay a few days and inject money into the local aboriginal economy and most visitors are in awe of the world's most famous monolith, especially at sunset and sunrise.

There is little point in visiting Ayers Rock unless it is climbed. Over seven million people have had the Ayers Rock climb experience over the last 60 years and there are still millions more who want to climb it. Some have died climbing Ayers Rock, that is only to be expected when people have the right to make up their own minds about their abilities and some of those older people who died at the peak were happy with their achievement.

Memorials to those who have died and monuments that mark the summit are to be removed from the Rock in an act of cultural vandalism yet it is argued that Ayers Rock is so culturally important that it is a no-go zone. The rules about preserving heritage in this National Park apply to both the aboriginal heritage and non-aboriginal heritage; both deserve respect. I managed to climb it before the chain and, some 40 years later, with the chain.

At the summit, one not only has a great view but is in awe of nature, is fulfilled and exhilarated after the effort and is dwarfed by the distances, colour, beauty, barrenness and the insignificance of what it is to be human. There is no doubt, as Marc Hendrickx argues, that in pre-historical times Ayers Rock was climbed and climbers experienced the same vista, exhilaration and feeling of insignificance. These feelings are what it is to be human.

And there is the rub. Ayers Rock, like so many stupendous features of nature, belongs to mankind. During historical times, there have been great changes. Previously, aboriginal people did not care less whether someone climbed Ayers Rock or not. However, the politics of aboriginal groups is ever changing and now a confected cultural conflict is proposing that climbing be stopped. Most tourists that come to Ayers Rock want to climb it. That is the purpose of the visit. If they can't come and climb it, they spend their money elsewhere. Does this mean that the aboriginal community is so well funded by "sit down money" that they can afford to destroy a viable business? Why is it that previous leaders of aboriginal groups proudly displayed their culture at Ayers Rock to tourists where the modern politicised leaders now claim that their culture must be kept secret?

One just can't help thinking that a proposed no-go zone at Ayers Rock has nothing to do with culture and has everything to do with political power, grant money and a misplaced sense of entitlement. Maybe a no-go zone is yet another symbol of the taking of the majority's freedoms by unelected minorities. Those who climb, bush walk and camp in nature are the true environmentalists and sympathisers of past cultures whereas those that deny these simple pleasures, be they Ayers Rock, Mt Warning or St Mary's Peak, are living in cocoons isolated from those that they allegedly serve.

Emeritus Professor Ian Plimer

"It is not the mountain we conquer but ourselves."

Edmund Hillary

INTRODUCTION

Ayers Rock has been around for tens of millions of years before humans first thought about climbing it, and it will be there long after the last climber has departed. The petty politics that govern access to its summit is an insult to its natural grandeur. Visitors to this remarkable wonder have the right to make up their own minds whether or not to partake in a simple exploration of nature without being harassed by transitory ideologies, strange religious beliefs or extreme political views. Everyone has the right to enjoy the natural world without being made to feel guilty.

Everyone who has completed the strenuous walk to the summit of Ayers Rock and exhilarated in the joy and wonder of the experience has a special tale to tell. Since well before the Ayers Rock-Mt Olga National Park was officially gazetted in 1958, to the present day, those wonderful stories have formed the foundation of the Park's success. Some of those stories may be found in this book: from the tales of those who first arrived at the Rock maybe as long as 35,000 years ago, to the current Traditional Custodians who likely arrived there about 500 years after the Great Pyramid of Giza was built, to the early European explorers and millions of Australian and International Tourists who came later. These tales of exhilaration, excitement, joy, wonder, fear and sometimes tragedy, form the core of the Park's reputation as an international tourist icon.

Since 2001, increasingly these breathtaking tales are being told in hushed tones. Undertaking the journey up that western climbing spur is now a bit like joining Chuck Palahniuk's Fight Club, the first rule; you don't talk about the Climb. The second rule, **You Do**

Not talk about the Climb! This is lest you be assaulted with accusations of racism, intolerance and ignorance, intended to make you feel embarrassed and ashamed. This is due to a blitzkrieg of propaganda against the Climb being launched by the Park Board with the support of weak-kneed, petty bureaucrats from Parks Australia and hypocritical politicians. As this book will demonstrate, pretty much everything the Board and Parks Australia say about the Climb is a myth.

In 1991 the Board and Parks Australia said the Climb was a sacred site, yet in the 1970s we were told by the Principal Owner of the Rock, Paddy Uluru, that the physical act of climbing was of no cultural interest. His brother, Toby Naninga, stated in a television interview in 1975 that aside from Warayuki, the men's initiation cave, and the nearby Ngaltawata Pole, tourists could go "anywhere else". In the 1940s Anangu man, Tiger Tjalkalyirri, guided tourists up the Rock and was filmed splashing about in rock pools and horsing about on the summit, enjoying their company.

Who are we to believe – members of a Board far removed from a traditional nomadic life, or those past elders who were more closely connected to their land and their law? Visitors are told that Tjukurpa, the local Aboriginal religion, is unchanging. It seems that either the Board are disrespecting or ignoring the old men's views about the Climb, or Tjukurpa is as arbitrary as any other religion. You can't have it both ways.

The Board and Parks Australia say the Climb is a safety risk, yet the statistics collated in this book indicate otherwise. Based on the total number of tourists, climbers and the 18 corroborated deaths (six falls and 12 from natural causes) that may be attributed to the Climb, climbing Ayers Rock is safer than many other tourist activities like the Grand Canyon and Great Barrier Reef. No one is talking about closing these other sites or activities due to the misfortune of unlucky, clumsy, ill-prepared or irresponsible tourists.

The Board, Parks Australia and sections of a biased media say climbers are disrespecting the Rock, but out of the seven million people who have climbed it they typically only identify two individuals: Sam Newman, an AFL legend who drove a golf ball into the desert from the summit, and a French exotic dancer, Alizee Sery, who took her clothes off on top to honour the locals. These are very minor indiscretions that have been blown out of all proportion. The fact is that the overwhelming majority of climbers approach the Climb with great care and respect, as they do any other natural attraction in any other National Park. While the Board exaggerates the environmental impact of the minor amount of human waste left on the Rock, they seem to ignore the environmental damage of visitors caught short or littering on the longer walk around the base, or on walks around The Olgas.

The Board and Parks Australia say that the proportion of people wanting to climb has dropped below 20%. Yet they don't tell you that the Climb is closed for 80% of the time denying the majority of visitors the opportunity to even make the choice. A statistical analysis presented in this book, using data obtained through Freedom of Information (FOI), demonstrates that when the Climb is fully open the proportion of visitors climbing is sometimes higher than 80% and averages 44%. This is a better reflection of visitor intentions than including statistics on those days when the Climb is completely closed.

The Board compares the Climb to Disneyland and say they want to close it down. In doing so they demonstrate they have never been anywhere near that place, and have probably never climbed their Rock, like their grandfathers did before them. What sort of miserable bastard would close down Disneyland anyway? The Climb, a demanding physical challenge, is far from the gimmicky attractions of Anaheim, yet these are exactly the sort of attractions the Board is permitting in the Park, ironically, ostensibly to replace the Climb that they want to ban. Unlike the Climb all these "attractions" come with a considerable price tag, a proportion of each ticket finding its way back to the Park Owners. Recent years have seen Segways getting approval to circle the Rock, skydivers to jump… some distance from it, and bikes to peddle or motor around it. The most recent addition is the SkyShip Uluru, a tethered helium-filled balloon stationed a whopping nine kilometres from the Rock that will only go up a third of its height. The Board doesn't seem to realise it is not closing down Disneyland, they are turning their home into a sideshow alley version of it.

Parks Australia and the Board say they are acting within the law in banning the Climb and effectively destroying its associated physical heritage: the summit monument, the chain and five memorial plaques at the base. Along with the act of climbing, the Board seems intent on removing all non-Anangu cultural heritage from the Rock. However, the lease agreement requires Parks Australia to preserve, maintain and manage cultural heritage in the Park to the highest international standards.

As this *Guide* will demonstrate, the activity of climbing the Rock and the physical objects associated with the Climb represent significant Australian cultural heritage, and Parks Australia have breached the lease in colluding with the Board in authorising their removal. In 2009 former shadow Environment Minister Greg Hunt warned that Big Brother was coming to the Rock. Sadly it seems his prediction is coming true.

This book explodes these myths and shows the Climb is something to be celebrated, preserved and maintained; not banned out of superstition and petty politics.

The *Guide* looks at the long history of climbing, and provides incontrovertible evidence that explodes the "We Never Climb" myth promulgated by the Board and Park Management. It outlines reasons for climbing that Parks Australia and the Board don't want you to know about. The *Guide* tells you everything you wanted to know about the Climb but were too afraid to ask.

This book does not contain much information about Aboriginal Mythology at Ayers Rock, for that the reader is referred to the excellent work of Charles Mountford including: *Brown Men and Red Sand, Ayers Rock: It's people, their belief and their art*, and *Nomads of the Australian Desert*. The importance of Mountford's ground-breaking anthropological research in central Australia has been understated, traduced and downplayed by the current Board of Management and Parks Australia who do not acknowledge his work in the recent Plans of Management or in Park literature. His stolid determination to obtain consistent stories from a wide variety of original sources, over decades, was once rewarded, but is now a piece of history and academic research that it seems the Board and Parks Australia would prefer be erased.

As well as being a simple guide to the world's most famous hill climb, this book is also a political communication, part parody and satire. Sadly, because the simple act of enjoying the natural world has been tainted by propaganda and political correctness, and been used as a political football since the early 1990s, people have lost sight of what is important. Facts have been muddied and history erased. Prior to politics getting in the way, Ayers Rock was overwhelmingly a fun place to visit. Hopefully by reading this book, coming to grips with the facts and undertaking the Climb we can all recapture the joy that has been lost for so long, and, just perhaps, this ridiculous ban can be overturned.

Every effort has been made to obtain correct factual information about the Climb and its history from original sources, but no-one is perfect. If you find something you disagree with or can provide more details and original material about matters herein, please contact the author via the publisher. Subsequent editions will be improved thanks to your input.

A comprehensive list of endnotes may be found at the end of the book. Year format used is day/month/year, "yBP" means years before present, and "Mya" means millions of years ago.

The Climb belongs to the world. It is an iconic walk that reveals views of nature that are among the most outstanding on the planet.

Its human history belongs to all of humanity not just a selfish few who want to keep their culture a secret from the world, and sadly even from each other. The proposed ban on climbing is an insult to past Traditional Custodians who climbed and the seven million visitors who have also climbed. Why would anyone want to destroy something that has provided inspiration to so many? What madness has taken hold of us, that in 21st century Australia such irrational actions could be contemplated let alone undertaken with such mild objections?

Ayers Rock Facts and Figures

The Park

Official name of the Rock	Uluru or Ayers Rock
	On 15 December 1993 the feature was officially dual named Ayers Rock / Uluru (where both names are equally as important and can be used either together or individually).
	Following a request from the NT Regional Tourism Association, on the 6 November 2002 the order of the dual names was officially changed to Uluru / Ayers Rock.
	It is also believed that the name Uluru itself is more directly associated with the water hole above Mutitjulu (Maggie Springs) than with the Rock as a whole.[1]
Date Park opened	Ayers Rock-Mount Olga National Park gazetted 20 February 1958, but available from early 1957 with Bill Harney appointed the first Ranger in March 1957.
	Declared under the Commonwealth *National Parks and Wildlife Conservation Act 1975* as the Uluru (Ayers Rock-Mount Olga) National Park on 24 May 1977, and name changed to Uluru-Kata Tjuta National Park in 1993.
Area	1,333.72 km²
Longest Serving Ranger	Derek Roff MBE 1968-1985.
Custodians	From 35,000 yBP, pre-Anangu peoples. From 4,000 yBP, Pitjantjatjara and Yankunytjatjara peoples. From 1901, Australians.
Ownership	Uluru-Kata Tjuta Aboriginal Land Trust from 26 October 1985.

Physiography

Height of Ayers Rock	863m (865m noted on Bronze Directional Plate at summit).
Height of Ayers Rock above Plain	348m above the surrounding plain.
Height of Mount Olga	1066m

Height of Mount Olga above Plain	545m, almost 200 metres higher than Ayers Rock
Summit coordinates	S25° 20.694' E131° 01.955'
Circumference	About 9.5km
Average height of sand dunes	10m

Weather

Highest maximum Temperature recorded	47.0 °C 31/12/1993 Yulara Airport (BOM station 15635)
Lowest minimum Temperature recorded	-5.0 °C 19/7/1976 Ayers Rock (BOM station 15527)
Highest wind speed recorded	118km/h 23/12/2009 Yulara (BOM station 15635)
Highest daily rainfall	133.6mm 7/3/1967 Ayers Rock (BOM station 15527)

Climbing

First Climbers	From circa 35,000 yBP: Pre-Anangu peoples
First Anangu Climbers	From about 4,000 yBP (after Dingo introduced into Australia)
First non-Anangu Climbers	20 July 1873, William Gosse and Kamran
First woman	Isabella Foy (?) 28 May 1936 (named in logsheet at cairn)
Second woman	Beryl Miles mid-1951 (possibly first woman climber)
Oldest climbers	Several reports of people in their 80s climbing.
Most climbs by one person	Howard Rust over 100 times[2] (the 100th on his 61st birthday).[3]
Most routes up	Graham Phillips – seven routes
Fastest up	Under 13 minutes
Fastest down	Under 11 minutes
Proportion of visitors climbing	Up to 80%, on days the Climb is fully open
Overall proportion of visitors climbing from total visitors 1958-2017	60%

Number of individual climbers 1958-2017	About 7 million
Length of Climb (Chain route)	About 1560m
Percent time closed (weather or culture)	77.3 % of the time in 2017
Chain installed	1964, completed 1976, extended in early 1980s
Summit marker	Erected William Gosse 1873
Summit Survey Cairn	Post and vanes: constructed 15/5/1958 (removed Dec. 1970)
Summit pedestal and Bronze Directional Monument	Constructed December 1970. Bronze map of Australia lost by early 1980s, Australian Coat of Arms lost early 2000s
Log books	20/05/1966-24/05/1986: 171 logbooks recording the names of about 1.3 million climbers. NT Archives Alice Springs
Deaths on the Rock (corroborated)	6 falls, 12 heart attacks on the Rock (as 2/8/2018) Parks Australia claim 37
Youngest Royal to Climb	Prince Hiro of Japan (14), August 1974
First "No climb" sign	1991
Cost of discouraging climbing	Estimated at over $70 million per annum[4]

Tourism

Total visitors 1958-2017	11.7 million
First Motel leases	Late 1950s
First Plane Arrival	Donald Mackay Expedition June 1930
Airstrip at Rock opened	20/4/1958 Eddie Connellan pilot
Ayers Rock Airport opened	6/6/1982, by Malcolm Fraser
First Vehicle access	Michael Terry: 1930, Foy Expedition: 1936
Graded Road opened	November 1948
Highest visitor numbers	2000-2001: 436101, (raw figures adjusted by 10% to allow for children
Yulara Resort opened	Operational late 1984

1

HISTORY OF CLIMBING AYERS ROCK

History of Climbing Part 1: 33,000BC to 1873AD

The proposed ban on climbing Ayers Rock will bring to an end a 35,000-year-old tradition of climbing that likely began when the first humans arrived at the foot of this grand rock outcrop.

The arrival of humans in the arid region of central Australia is dated at about 35-29,000 yBP[5] from two cave shelters: Puritjarra and Kulpi Mara in the Western MacDonnell Ranges about 160km north of Ayers Rock, 300km west of Alice Springs. Radiocarbon dating of deposits at these sites shows pulses of activity between about 35,000 years ago and the near present.[6] No dating has been undertaken at Ayers Rock, but it is reasonable to assume that the area probably saw intermittent habitation by humans around the same time, give or take a millennia or two. Given the universality of human curiosity and the strategic importance of Ayers Rock as a water resource and vantage point, it is likely that the first climbers of the Rock were among these first arrivals. From their lookout they would have seen the end of the megafauna, and seen off the ice age that ended the march of the sand dunes across the surrounding desert plains. Over the ages these early visitors left little to show for their presence except for a few undated petroglyphs on the northern face of the Rock near the Tjinindi Rockhole[7] indecipherable to both archaeologists and the current traditional custodians. Similar petroglyphs abound in central and northern Australia. Climate in central Australia during the last ice age was much drier than now. Human habitation at the Rock in the presence of decade long mega droughts must have been an even more difficult proposition compared to today and it's likely that the Rock was only visited during wetter intervals in that period.

The first real evidence for climbing can be found in the oral history of the current custodians; in the creation myths and legends of the Yankunytjatjara, Pitjantjatjara and Ngaanyatjarra peoples. Collectively these Central Desert groups are known as the Anangu. According to Anangu creation myths the route of the current climb was that taken by the ancestral Mala men (Hare Wallaby men) on their arrival at Ayers Rock.[8] The legends likely portray events in the early history of the Anangu experience at the Rock. Perhaps they displaced other people, or perhaps the area was abandoned when they arrived. Without a reliable water source settlement at the Rock was always transitory and dependent on the availability of water in pools at the base and the summit. During long periods of drought it is unlikely that the Rock was occupied. Permanent settlement only became an option following the construction of water bores in the 1950s. Related creation myths provide a means to constrain the timing of this prehistoric arrival. According to Anangu oral legends Kulpunya (a "spirit dingo") destroyed most of the Mala men and their families during the creation of Ayers Rock.[9] The dingo was introduced into Australia by Asian seafarers about 4,000 years ago.[10] Based on this, the Anangu would have likely arrived at Ayers Rock and developed their mythology sometime after 4,000yBP, give or take several hundred years (see also Griffin 2002[11]). Hence the "Mala men" may have made their first ascent on the Rock 500 years or so after the Great Pyramid of Giza was constructed.

In the time after the arrival of the "Mala men" up to European exploration and settlement in the late 19th century the current custodians and various other tribes regularly visited the summit for initiation rites, *inma* ceremonies, and likely took advantage of the view to look out for game and provide a warning of the approach of rival clans. Anthropologist Charles Mountford (1965) recounts a number of Anangu myths and legends on the summit of Ayers Rock that indicate excellent knowledge of summit features that would have required regular visits to the top of the Rock over generations. These include the tale of the lizard man Linga who murdered a Kunia

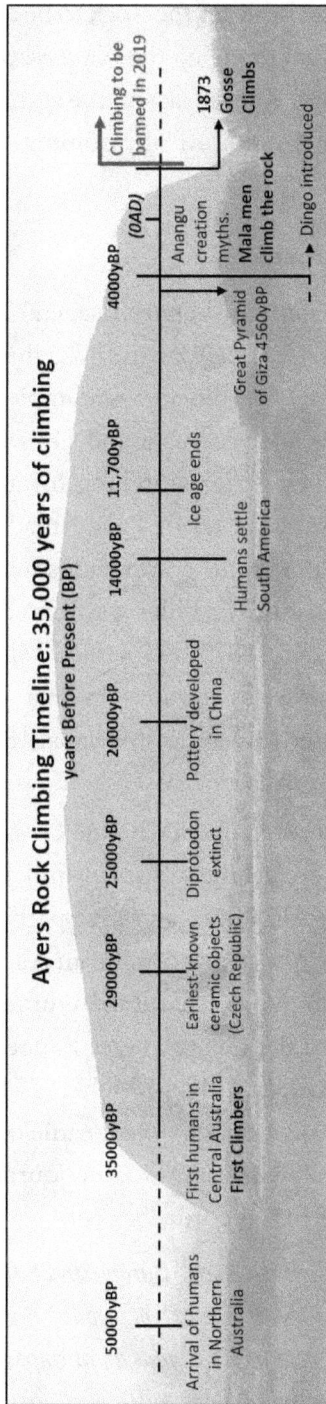

Ayers Rock Climbing Timeline: 35,000 years of climbing
years Before Present (BP)

50000yBP	35000yBP	29000yBP	25000yBP	20000yBP	14000yBP	11,700yBP	4000yBP	(0AD)	1873

- Arrival of humans in Northern Australia
- First humans in Central Australia **First Climbers**
- Earliest-known ceramic objects (Czech Republic)
- Diprotodon extinct
- Pottery developed in China
- Humans settle South America
- Ice age ends
- Great Pyramid of Giza 4560yBP
- Anangu creation myths. **Mala men climb the rock**
- Dingo introduced
- **Gosse Climbs**
- Climbing to be banned in 2019

Figure 1: Ayers Rock climbing timeline 50,000yBP-Present[12]

woman and dragged her body up the Rock to be eaten, gutters formed by the Hare Wallaby Men dragging the *Ngaltawata* pole to safety from the attacking spirit dingo Kulpunya, and the stories about the *"Uluru"* waterhole on the eastern side of the summit plateau, home of a mythical serpent.

First European climber

There was some competition between several European explorers to survey the region. On 22 August 1872, the same day the first transmission was sent over the newly completed Overland Telegraph line connecting Port Augusta and Darwin, Ernest Giles set out west from Chambers Pillar searching for an overland route to Perth and got as far as the shores of Lake Amadeus. Unable to cross the boggy salt lake and with his companions wanting to return having had enough of the journey, Giles was forced to go back, ending up in Adelaide in January 1873. He named Mt Olga and the Lake during this expedition. Giles set out again on a second expedition to trace a route into Western Australia in August 1873, but he would be beaten to the Rock by William Gosse.

In April 1873 surveyor William Christie Gosse, commissioned by the South Australian government also to find a route from central Australia to Perth, left Alice Springs with a party of four white men, three Afghans, and an Aboriginal boy. From Alice Springs the group initially headed north to Mt Leichardt then turned south, passing Mt Liebig, Glen Edith and the George Giles Range. The party successfully crossed Lake Amadeus and on 19 July 1873 they arrived at Ayers Rock.[13] Gosse, accompanied by Afghan cameleer Kamran, climbed up bare-footed the following day. Gosse's journal entry documents the excitement of this first recorded ascent:

Sunday, July 20 — Ayers Rock. Barometer 28.07 in., wind east. I rode round the foot of rock in search of a place to ascend; found a waterhole on south side, near which I made an attempt to reach the top, but found it hopeless. Continued along to the west, and discovered a strong spring coming from the centre of the rock, and pouring down some very steep gullies into a large deep hole at the foot of rock. This I have named Maggie's Spring. Seeing a spur less abrupt than the rest of the rock, I left the camels here, and after walking and scrambling two miles barefooted, over sharp rocks, succeeded in reaching the summit, and had a view that repaid me for my trouble—Kamran accompanied me. The top is covered with small holes in the rock, varying in size from two to twelve feet diameter, all partly filled with water. Mount Olga is about twenty miles west. Some low ranges and ridges west-north-west, one of which I think must be McNicol's Range; part of lake visible, bearing north Mount Conner 96°, and high ranges south-east, south, and south-west, with sandhills between. The one south-east I have named after His Excellency Governor Musgrave; and a high point in same, bearing 141°, Mount Woodroffe, after the Surveyor-General. This is a high mass of granite, the surface of which has been honeycombed, and is decomposing, 1,100 feet above surrounding country, two miles in length (east and west), and one mile wide, rising abruptly from the plain. How I envied Kamran his hard feet; he seemed to enjoy the walking about with bare feet, while mine were all in blisters, and it was as much as I could do to stand: the soil around the rock is rich and black. This seems to be a favourite resort of the natives in the wet season, judging from the numerous camps in every cave. These caves are formed by large pieces breaking off the main rock and falling to the foot. The blacks make holes under them, and the heat of their fires causes the rock to shell off, forming large arches. They amuse themselves covering these with all sorts of devices—some of snakes, very cleverly done, others of two hearts joined together; and in one I noticed a drawing of a creek with an emu track going along the centre. I shall have more time to examine these when the main camp is here. This rock is certainly the most wonderful natural feature I have ever seen. What a grand sight this must present in the wet season; waterfalls in every direction. I shall start back, tomorrow, and trust to finding some water between here and King's Creek, which is now eighty-four miles distant.

Gosse returned to King's Creek the following day, returning on 28 July to establish a depot near the waterhole he named Maggie

Springs (now called Mutitjulu waterhole) remarking: *Depot No. 8, Ayers Rock. Barometer 28 in.; latitude 25° 21' 28" south. This rock appears more wonderful every time I look at it, and I may say it is a sight worth riding over eighty-four miles of spinifex sandhills to see.*

Gosse also makes the first written account of viewing waterfalls cascading off the Rock after rain. *Friday, August I.—Depot No. 8, Ayers Rock. Barometer, 27.87 in.; wind N.W. The rock presented a grand appearance this morning; close to our camp was a waterfall about 200 feet high, the water coming down in one sheet of foam.*

He left the Rock on 8 August to continue his explorations to the west, his tracks at Mt Olga being found by Ernest Giles only a month later. Gosse's expedition failed to reach Perth, making it as far west as Mount Whitby, about 200km west of the South Australian border before turning back to the overland telegraph line near Oodnadatta via the Musgrave Ranges and the Alberga River.

The remarkable reaction to the Rock and the Climb first recorded by Gosse has been felt by millions of visitors over the last 145 years and, no doubt, by those that climbed before him. The tradition of climbing the Rock is not unique to Europeans but has been practised by all visitors since humans first arrived in the region some 35,000 years ago.

Figure 2: Sketch of Ayers Rock from Gosse's journal

History of Climbing Part 2: 1873-1958

Following Gosse's ascent in 1873 several explorers and scientific expeditions visited the Ayers Rock area. Ernest Giles in 1873-74, William Henry Tietkens in 1889, the Horn Expedition in 1894, Maurice and Murray in 1902 and F.H. Hunn in 1906.[14] There seems to be lull in reported visitors over the course of the early 1900s, with renewed but sporadic reports in the 1920s and 1930s. Visitors in the 1940s were assisted by the first tourist guide: Anangu man Tiger Tjalkalyirri. Tourism to the Rock was opened up in the late 1940s with a graded track constructed in 1948; Arthur Groom was perhaps the last to get there by camel in 1947. Tour bus services began in the early 1950s led by local tourism entrepreneur Len Tuit who should be given credit for opening the Rock up to travellers.

With improved access the number of visitors climbing the Rock began to swell. In 1958, the area that is now the park was excised from the Petermann Aboriginal Reserve to be managed by the Northern Territory Reserves Board as the Ayers Rock-Mount Olga National Park. The first ranger was legendary Central Australian figure Bill Harney appointed prior to the official declaration of the Park in March 1957.[15] By the end of 1950 there were only about 70 names recorded in slips of paper stored in glass jars at the small pile of stones that marked the summit, but by the end of this decade there would be hundreds. Formal counting of visitation began in 1958 when there were 2296 visitors to the park.

Ernest Giles, 1874

On his second expedition into the region Ernest Giles camped at Ayers Rock along with William Tietkins and the rest of his party on 9 June 1874, almost a year after Gosse had visited and scaled it. Giles remarked on the climbing route but does not mention climbing: *Its appearance and outline is most imposing, for it is simply a mammoth monolith that rises out of the sandy desert soil around, and stands with a per-*

pendicular and totally inaccessible face at all points, except one slope near the north-west end, and that at least is but a precarious climbing ground to a height of more than 1100 feet.

William Henry Tietkens, 1889

William Henry Tietkens had accompanied Giles in earlier explorations of central Australia. In 1889 he led an expedition into Central Australia on behalf of the Central Australian Exploring and Prospecting Association. Like Giles he mentions the climbing route used by Gosse but does not mention climbing:[16]

> *Tuesday July 9th – Camp 62… Started away at 8 am., Ayers Rock being five or six miles away. Travelled over sandy flats covered with mulga scrub, passing occasional spinifex sandhills, and soon reached the grass flats at the foot of the rock. Passed around the south base of this mountain of unbroken, unfractured stone. At 11:30 I unsaddled upon the spot where I camped fifteen years ago, when here with Giles. After dinner we went out to explore and admire this wonder in solid granite. Many and varied are the wonderful shapes it assumes. In one or two places large caves are to be found near the foot, and these in a measure spoil the otherwise regular and graceful lines chiselled by nature upon its face. The rock formation is a coarse-grained grey granite, the surface all over bearing a reddish color from exposure to the elements, smooth as glass, and almost polished, It appears to me to be quite inaccessible, except at the one point where Mr. Gosse and his companions made their ascent.*

Like many other early visitors Tietkins' description of the lithology of the Rock is incorrect. While it's likely derived mainly from a granite rich source, the Rock itself is a very coarse Arkose Sandstone (see Chapter 2). Tietkins took the first photographs of Ayers Rock.

Horn Expedition

The Horn Expedition of 1894 was a significant scientific undertaking equipped and sponsored by William Austin Horn, a wealthy pastoralist and mining magnate, who accompanied the expedition in

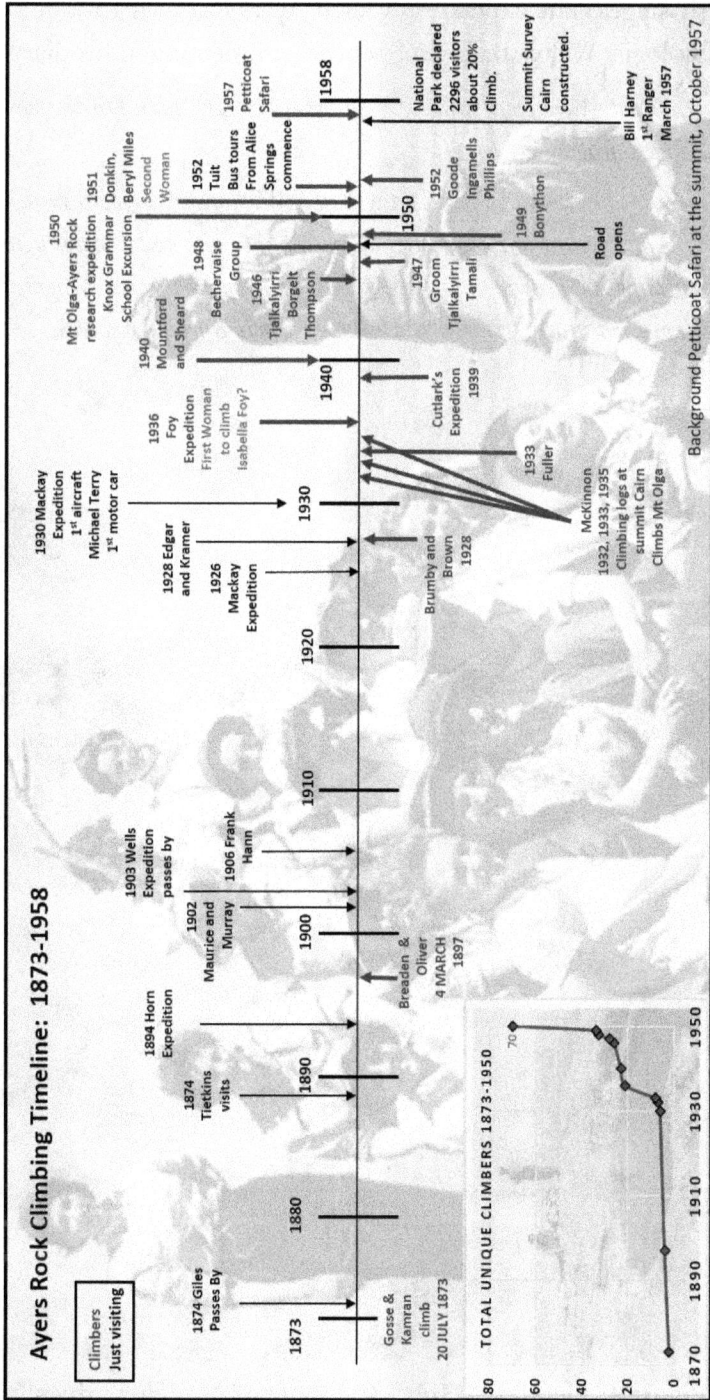

Ayers Rock Climbing Timeline: 1873-1958

Climbers Just visiting

1873 — 1880 — 1890 — 1900 — 1910 — 1920 — 1930 — 1940 — 1950 — 1958

1873
1874 Giles Passes By
Gosse & Kamran climb 20 JULY 1873
1874 Tietkins visits
1894 Horn Expedition
1897 Breaden & Oliver 4 MARCH 1897
1902 Maurice and Murray
1903 Wells Expedition passes by
1906 Frank Hann
1926 Mackay Expedition
1928 Edgar and Kramer
Brumby and Brown 1928
1930 Mackay Expedition 1st aircraft Michael Terry 1st motor car
McKinnon 1932, 1933, 1935 Climbing logs at summit Cairn Climbs Mt Olga
1933 Fuller
1936 Foy Expedition First Woman to climb Isabella Foy?
Cutlark's Expedition 1939
1940 Mountford and Sheard
1946 Tjalkalyirri Borgelt Thompson
1947 Groom Tjalkalyirri Tamali
1948 Bechervaise Group
1949 Bonython
Road opens
1950 Mt Olga-Ayers Rock research expedition Knox Grammar School Excursion
1951 Donkin, Beryl Miles Second Woman
1952 Tuit Bus tours From Alice Springs commence
1952 Goode Ingamells Phillips
1957 Petticoat Safari
National Park declared 2296 visitors about 20% Climb.
Summit Survey Cairn constructed.
Bill Harney 1st Ranger March 1957

Background Petticoat Safari at the summit, October 1957

TOTAL UNIQUE CLIMBERS 1873-1950

0 — 20 — 40 — 60 — 80

1870 — 1890 — 1910 — 1930 — 1950

Figure 3: Ayers Rock climbing timeline 1873-1958[17]

9

its early stages. The expedition visited Ayers Rock on 15 June 1894, with biologist Walter Baldwin Spencer commenting in his diary:

In camp at Ayers Rock Watt & self went out early round rock. Copied black drawings. Photod

Sides very precipitous. Possible ascent at S.W. 'corner'. Smooth surface with scales peeling off. Big blocks all round base with hollow caves with crow dung & drawings. Fig trees. Mulga Kangaroo grass. Magpie's spring – not nearly so large as when Cowle here 8 weeks before.[18]

Figure 4: Tietkins' 11 July 1889 photo of Maggie Springs[19]
(State Library NSW Collection)

Figure 5: Ayers Rock by Baldwin Spencer, 1894[20]

Spencer attempted to reach the summit via the climbing spur but turned back at a height of 200 feet (a little bit above Chicken Rock). He noted in *Across Australia* (1912):

> *Just at sunset we made an attempt to climb, but were quite content to stop when we reached a height of about 200 feet. The surface was so smooth and steep that we could only climb by means of clinging on with our fingers to little projecting flakes of rock. Every now and then there was a hump, on the upper surface of which the incline was less steep than elsewhere, and on one of these we came to rest, quite satisfied that the climb to the top was not worth the attendant risks, as the slightest slip would have been fatal. We were, however, amply repaid for the trouble of our short but uncomfortable climb. Beneath us, the sandy plain was dotted over with thin scrub and, away in the distance, it was crossed by dark lines where, mile after mile, the thick mulga scrub stretched across. The level line of the horizon was only broken by the great, dome-shaped masses of Mt. Olga, behind which the sun was setting and, against the rich orange of the western sky, its purple masses stood out in strong relief. Our camp fire began to show out clearly in the dark chasm beneath us and, to complete the scene, we saw a family of the wild, sand-hill natives making their way, led by our own black boy, round the base of the mountain towards our camp.*[21]

**Figure 6: First photo of the Olgas by Baldwin Spencer, 1894, and
first photo taken from Ayers Rock**[22]

Spencer refers to Charles Ernest Cowle, the policeman at Illamurta Police Station. Cowle escorted the Horn Scientific Exploring Expedition from Illamurta, across the large expanse of Lake
Amadeus to Ayers Rock, and back. There is no indication Cowle
climbed the Rock. The leader of the family Spencer saw on the
Rock, *Lungkartitukukana*, is thought to be have been Paddy Uluru's
father.[23] Paddy Uluru was acknowledged as the principal owner of
the Rock in the 1970s.

On 19 June 1894 Spencer took the first photo of the Rock in its
entirety, showing the west face with the climbing spur prominent in the
centre right, probably from the vicinity of the current sunset viewing
area. His photo of the Olgas taken from low down on the climbing
spur was the first of those round-headed peaks, taken from the Rock.

Breaden and Oliver 1897

Allan Breaden and David (Billy) Oliver were the second to record
their climb to the summit and the first to establish the practice of
leaving a record of the achievement at the summit cairn.[24] Breaden's

journal entry for 4 March 1897 indicates the Climb was not very difficult, but he and Oliver had sore feet for days afterwards:

March 4, 1897. Camped at Ayers Rock. Walked around rock and examined water and the chance of climbing to the top which Oliver and I did in the afternoon. The day happened to be very calm, and the ascent was not very difficult as far as the climbing was concerned, but we got very footsore, hardly able to walk for days after. It would not be safe to climb (the rock) on a windy day as any slip would cost a man his life. There is nothing to save him, and he would simply roll until he reached the plain below.

The Rock is fearfully scarred with lightning all along the top. Gosse, the explorer, is the only other white man who has been on the top. We found his small pile of stones about the centre, but no other indication that a white man had ever been there. We added a few stones to the pile and left two wax vesta boxes (tins) with names and date thereon. One is unable to take one's eyes off this wonderful natural feature. Its height is, I believe, about 1100 feet above the level of the plain, and in most places it is quite perpendicular and absolutely bare of vegetation.

Breaden is reported[25] as likely the first, along with Bob Coulthard, to undertake a west to east traverse of the Rock, but the date is not known: *Ayers Rock, again in the recent news. Mr. McKenzie told me that he knew two bushmen, Bob Coulthard[26] and Alan Breaden, who scaled up the western side and left their names there in a baking powder tin, and then carefully tobogganing down the east side.* (See Chapter 4 for possible route taken).

Allan Breaden was born at Booboorowie Station in northern South Australia in 1854 and died in 1943. He was a stockman and explorer and lived in Central Australia for almost all his life. The Breaden family owned Henbury Station for much of the early 1900s.

Maurice and Murray, 1902

The *Adelaide Advertiser* (29/11/1902) reports on the exploration of Central Australia by R. T. Maurice and W. B. Murray: *The party consisted of Mr. R. T. Maurice (leader), Mr. W. R. Murray (surveyor),*

Figure 7: Allan Breaden, the pioneer of the Finke district standing beside his brother Joe's memorial to the pioneers of the Finke at Henbury Station[27] (State Library of South Australia B 38778)

H. Hauscheldt (cook and general hand), Khasta Khan (Afghan camel driver), Munjena, Yarrie, and Peter (the three black boys Mr. Maurice brought to Adelaide on his last visit), two other black boys, 14 camels, and Mr. Maurice's Airedale dog Jack.[28]

About Ayers Rock, Murray reports: *Arriving at that extraordinary feature, Ayers's Rock, which was first seen by Giles and named by Gosse after the late Sir Henry Ayers, we were enabled, after much searching, to get a drink for the camels and to fill the canteen. Mr. Murray attempted to ascend the rock by crawling backwards, but owing to the gale which was blowing he was obliged to give it up. We could well understand Mr Tietkins saying that as the rock had already been ascended he did not see the necessity of it being done again. It was at the western-side of the rock that we found splendid aboriginal drawings, especially the snake—an excellent drawing mentioned by Gosse and beautifully reproduced by Mr. Murray.*[29]

Wells' Expedition, 1903

The South Australian Government North West expedition (Wells Prospecting Expedition)[30] of March-November 1903[31] passed through the area. Herbert Basedow (1914) records a detour to Ayers Rock by F. R. George during this expedition, but it is not known that George or members of his party climbed.[32]

Hann, 1906

Prospector Frank Hann is reported[33] to have passed by Ayers Rock in 1906 coming from Laverton in Western Australia to Oodnadatta. There are no reports that he climbed it.

There seems to have been a lull in reports of exploration activity prior, during and immediately after the First World War.

Mackay Expedition, 1926

In 1926 the Mackay Expedition,[34] led and financed by wealthy pastoralist Donald Mackay,[35] explored the area between Oodnadatta and the Petermann Ranges.[36] Mackay was joined by anthropologist, geologist, explorer and medical practitioner Dr Herbert Basedow.[37] Basedow engaged the assistance of Frank Feast and Bert Oliffe, each of whom had been on two earlier trips with Basedow. At Oodnadatta three Aboriginal assistants were also engaged — Sambo, Ronald and Jack, all Kaiditj men. To spell the camels the party walked almost two-fifths of the 1300 miles (2092 km) covered.

Mackay and Basedow reached Ayers Rock on 19 June 1926. There are no reports that they climbed. Basedow described the Ayers Rock area:[38]

Figure 8: Members of the Mackay Expedition. Herbert Basedow tugging the shutter release cord to take the selfie, 1926.[39] Camels in background the Landcruisers of their day. How times have changed! (National Museum of Australia)

When aeroplaning becomes popular in Australia, not only Australians, but people from abroad will be drawn to Central Australia to see the colossal wonders of Nature in that, at present, remote territory. I have been all over the world; and have never seen anything to equal it. Ayers's Rock is a wonderful formation about 1200 feet high rising suddenly in the flat country. If it were in America it would be world-boomed. I never imagined such scenery existed. That so little is known of this part of our continent is due only to the tragic misfortunes of members of other expeditions, beginning with that of Giles, in 1876. The natives were always a danger. I certainly will go back by aeroplane sometime in the future to view again the marvellous scenery.

Edgar and Kramer, 1928

In 1928 the area was visited by J. Huston Edgar and E. E. Kramer[40] at the request of the Aboriginal Friends Association. Their objective:

... ascertaining the number of natives in the great reserves set apart for aborigines in South and Central - Australia; information about their food and water supplies, and the possibility of founding missionary stations.

Figure 9: Map showing Edgar and Kramer's Route from *The Observer Newspaper*, Saturday, 29 September 1928 (National Library of Australia Trove website)

Edgar and Kramer spent two months in the region passing by Ayers Rock and Mt Olga. There is no record of them climbing the Rock. They proposed future missionary stations at Ayers Rock and Mount Olga but these never progressed.

Brumby and Brown, 1928

Brumby and Brown are mentioned climbing the Rock in Michael Terry's 1932 book, *Untold Miles: Three gold-hunting expeditions amongst the picturesque borderland ranges of Central Australia.*[41]

As a young man Allan Ferguson Brumby worked as a dogger,

17

collecting Dingo scalps for the bounty offered by the South Australian Government.[42] Terry notes that Brumby and his mate T. H. ('Ginger') Brown climbed in 1928. He also indicates a couple of others had climbed by this time.

There is also mention of a "Brown and Brumley" climbing Ayers Rock in 1930, in a 1932 newspaper report in *The Queenslander*.[43] Under a photograph of Constable W. McKinnon the caption reads: *Constable W. McKinnon on the summit of Ayers' Rock. So far as is known the only other persons who have climbed this rock were Gosse the discoverer, about 1871, Alan Breadon in 1897, and Brown and Brumley in 1930.*

This is almost certainly a reference to the 1928 visit by Brumby and Brown.

Mackay expedition, 1930

Donald Mackay returned to Ayers Rock in 1930 by plane, as part of an aerial expedition he funded to survey remote parts of central Australia, landing on Tuesday, 17 June. The planes, *Love Bird* and *Diamond Bird*, were the first to be landed at Ayers Rock. Along with Mackay, the group included pilots Frank Neale and Bert Hussey, radio operator Howard Kingsley Love, navigator Commander H. T. Bennett and Philip Crosbie Morrison M.Sc., a young reporter working for the Melbourne *Argus*.[44] There are no reports of the party climbing Ayers Rock but they did walk around the base.[45]

Michael Terry – first vehicle at the Rock, 1930

Prospector Michael Terry led a number prospecting parties into Central Australia. In late 1930 he drove a Morris Truck to and around Ayers Rock and beyond, supported by a camel train. Terry's photos are a wonderful record of the area at the time. These is no indication he climbed the Rock. During their visit they found a camera lost by the Mackay Party a few months earlier.[46]

Figure 10: *Love Bird* at Ayers Rock in June 1930.[47] Photo E.A Crome collection of photographs on aviation (National Library of Australia)

Figure 11: Michael Terry's Morris Commercial truck in front of eroded caves, at Ayers Rock,[48] 1930 (National Library of Australia)

Constable William McKinnon

William McKinnon was appointed Constable with the Northern Territory Police Force in June 1931. He climbed the Rock a number of times, twice in early 1932, then again in 1933 and 1935.[49] In 1933 he replaced a match box tin left by Allan Breaden at the cairn with a glass jar. This held the names of climbers up to the early 1950s. The glass jar at the summit cairn purportedly recorded climbs by McKinnon on 7/3/1931 and 19/2/1932.[50] The 1931 entry is in error or due to a misinterpreted character as McKinnon did not commence work in the Territory until June 1931[51] and commenced his first patrol in October 1931.[52] His own photo record of his climbs of Ayers Rock notes climbs in 1932 (19/2/1932 and 7/3/1932), and again in 1933 and 1935.[53] The 1932 dates in the summit log were recorded during a four month patrol to Ayers Rock, Mt Olga and the Petermann Ranges that left Alice Springs on 18 January 1932.[54] It seems McKinnon climbed on the way out in February and then again on returning back past the Rock in March. He was also the first man reported to have climbed Mt Olga[55] on 21 February 1932.

McKinnon accompanied scientist Hedley Herbert Finlayson on an expedition to collect scientific specimens[56] in the Ayers Rock region and describes McKinnon's ascent (likely 1932):

> *During our first day at the rock, Mackinnon (sic) decided that the honor of the Commonwealth Police demanded that he should attempt to scale it. It is a difficult climb, and only four white men before him have succeeded, the ascent being made at the only practicable point, which is on the west side, where a buttress slopes down at a rather gentler grade than elsewhere. As my time was now short, Liddle and I were forced to leave him, with his boy, to win his Alpine laurels, while we pushed on for Mount Olga. But we learned subsequently that he had triumphed and, the day having been very calm, had reached the top in 40 minutes.*

In 1934,[57] following the murder of an aboriginal man near Mount Conner by two others for revealing ritual secrets to his wife,[58] Constable McKinnon tracked the suspects to Ayers Rock where he shot and killed Yokunnunna, the brother of Paddy Uluru. A subsequent inquiry found that McKinnon had acted legally, but that the shooting was not warranted.[59] The finding caused some controversy at the time,[60] and remains a source of grief for Traditional Custodians.

Figure 12: Photo of McKinnon standing at the cairn taken by self-timer in 1932[61]

The Climbers Log, 1932-1950

The glass jar placed by McKinnon contained slips of paper recording the names of those that climbed after him. It seems to have lasted until about 1958, supplemented perhaps by other containers. On 15 May 1958, the Australian Division of National Mapping replaced the small stone pile at the summit with a larger stone cairn and trigonometric survey marker with pole and vanes as part of the National mapping program. It's not clear what happened to the glass jars or other storage containers that may have been present and the notes they contained. Inquiries with Parks Australia have not yielded any success in locating these historical cultural artefacts.

Arthur Groom made a record of the list of names during his visit in 1947 (see below). The list was also recorded by members of the University of Melbourne Mt Olga-Ayers Rock research expedition in August 1950, and by the Knox Grammar School expedition in September 1950. These lists are reproduced below.

There are some differences between the three lists, with some names omitted from both, some names spelt differently, and some dates not matching other records. As indicated, the date of McKinnon's first ascent was 19/2/1932 as he notes in his own photos. Based on the list and factoring the 27 boys and teachers from Knox Grammar who climbed at the end 1950, from 1873 to 1950 at least 70 visitors are known to have climbed the Rock. It is likely that local Aboriginal groups also climbed during that period but there is no record of their ascents.

Mt Olga-Ayers Rock research expedition,[62] 15 August 1950

4/3/1931 W. McKinnon[63]

19/2/1932 W. McKinnon

1933 F. Walker

21/7/1933 H. Fuller

28/5/1936 H.V. Foy, Bill Foy, Mrs Foy, Tom McFadden, Denis Haycroft, Stan Tolhurst, Rupert Kuthner, Gre-----, Kurt----, Bill Mor----, S. Mulladad, Bob Buck

4/11/1939 V. Dumas, Billy Goat Mick, F. Clunes, F. Bailey (Cuttark's Expedition)

6/8/1940 C.P. Mountford

7/8/1940 L.E. Sheard

30/6/1946 C. Thompson

4/9/1947 Arthur Groome, Tamali, Talkajiori (sic- Tiger Tjalkalyirri)

23/9/1948 J.M. Bechervaise, Paddy de Conley, Sidney Stanes, Arthur Simpson, Ian Parkin

15/8/1949 C.W. Bonsthon (Bonython?)

15/8/1950 D.S. Kemsley, John Nicholas, Bob Smith, Leo Clarebrough, Alec Ewen, Alan Moore, Brent Greenhill, Alan Lofthouse, Trevor Broom *(Mt Olga-Ayers Rock research expedition)*

Knox Grammar School Expedition List,[64] 7 September 1950

7/3/1931 W. McKinnon *(see note above)*

19/2/1932 W. McKinnon

21/7/1933 H. Fuller, T.I. Whalker

28/5/1936 The Foy Expedition which included Bob Buck

4/11/1939 Cutlack's Expedition including V. Dumas, F. Clune, E. Bails

7/8/1940 C.P. Mountford, L.E. Sheard

14/8/1940 C.P. Mountford, L.E. Sheard

30/6/1946 L.A. Borgelt, Cliff Thompson, Tiger, Metingerie

21/4/1947 Arthur Groom

23/9/1948 J.M. Bechervaise, DuConnelly, S. Staines, Simpson, Parker

1949 Kimber, Ross, Bonython

6/5/1950 Malcolm R. Senior

14/8/1950 The Melbourne University Expedition

7/9/1950 Knox Grammar School Expedition: A.W. Briggs, A.C. Brown, W. Bryden, V.F.O. Francis, G.R.W. Latham, R. Millet, B. Rhys-Jones, D. Patten, M. Hughes, E.K. Chaffer, I. Brown, B. Piper, D. Grainger, B. Ross, W. Graham, I. Macfarlane, M. Lees, I. Quinlan, J. Bannigan, J. Neave, J. Williams, J. Young, J. Stranger, J. Graham, J Laurie, J.A. Walker, J. Schroder.

Noteworthy climbers to add their names to the glass jar at the summit between 1936 and 1948 include:

- The Foy Family (28/5/1936)
    - First family to climb, first woman recorded in logs at the summit cairn (Mrs Isabella Foy) though some doubts have been expressed as to whether all the people named on that list actually climbed.
- Cutlack's expedition (November 1939)
    - Frank Clune, V. Dumas, Billy Goat Mick, and F. Bailey
- Charles Mountford and Lauri Sheard (August 1940)
    - Recorded Aboriginal stories of summit features with Aboriginal guides
- Lou Borgelt, Cliff Thompson, Tiger Tjalkalyirri and Mitjenkeri Mick (30/6/1946)
    - First colour movie footage of the Climb and the summit!
- Arthur Groom, Tiger Tjalkalyirri and Tamali (4/9/1947)
    - Last one to visit by camel
- John Béchervaise (23/9/1948)
    - Second to climb Mt Olga (5/9/1948)

The Foy Family, 28 May 1936

Hugh Victor Foy was a director of Mark Foy Ltd, the well-known Sydney department store. In 1936 he set out with his second wife Isabella and son William on an expedition to find Lasseter's Reef.[66] They passed Ayers Rock on the way west. The glass jar at the summit contains a list of names of *all* the members of the Foy party written it seems in the same hand, except for Bob Buck's name at the end. An image of the sheet is captured in Lou Borgelt's 1946 movie. As best we can make out it lists names on the left hand side, titles in the centre and the place names Sydney and Alice Springs on the far right for the grouped names in the list. Text reads from top to bottom: H.V. Foy - Organiser(?), Mrs Foy - Passenger, Bill Foy - Passenger,

Tom McFadden - Driver, Denis Hayworth - Driver, Stan Tolhurst and Rupert Kuthner – Cameramen; all these from Sydney. Then from Alice Springs: Gus Schaller - Driver, Kurt Johannsen – Driver(?), Bill Morgan - *unreadable*, Sydney Walker - Guide, S. Mulladad - Cameleer, Bob Buck - Bushman. Doubts have been expressed (see Mountford, 1949;[67] and Latham, 1951[68]) as to whether all members of the group actually climbed. No photos were located showing the party on the summit.[69] If they did climb, then the Foy's were the first family to climb, and Mrs Foy would have been the first non-aboriginal women to scale the Rock. As the Anangu did not allow women to climb, and the climbing culture of pre-Anangu people is not known, she may have been the first woman. Her name written as "Mrs Foy" is the first woman's name to appear in the summit log.

While in the area of Lasseter's Reef, Foy produced a movie "Phantom Gold"[70] that documented the expedition with re-enactments of Lasseter's life. Footage includes scenes of the Rock

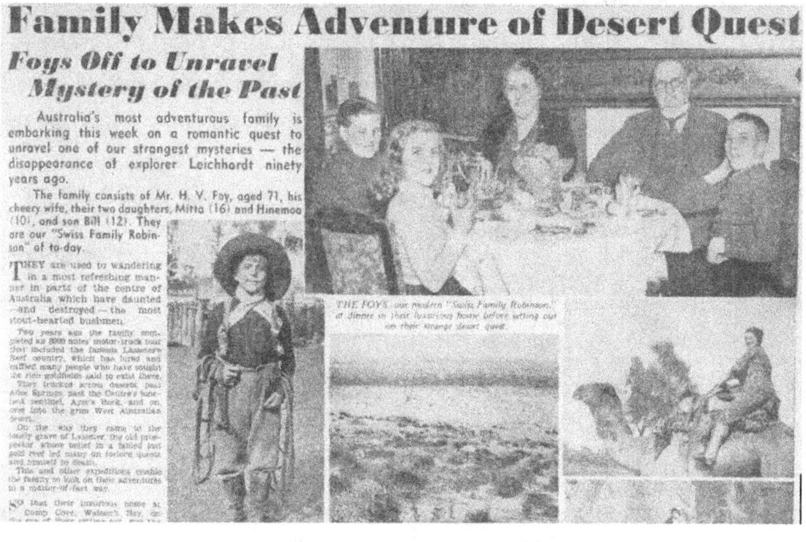

Family Makes Adventure of Desert Quest

Foys Off to Unravel Mystery of the Past

Australia's most adventurous family is embarking this week on a romantic quest to unravel one of our strangest mysteries — the disappearance of explorer Leichhardt ninety years ago.

The family consists of Mr. H. V. Foy, aged 71, his cheery wife, their two daughters, Mitta (16) and Hinemoa (10), and son Bill (12). They are our "Swiss Family Robinson" of to-day.

THEY are used to wandering in a most refreshing manner in parts of the centre of Australia which have daunted—and destroyed—the most stout-hearted bushmen.

Figure 13 Story about the Foy Family a modern day "Swiss Family Robinson" in the Australian Women's Weekly, 2 July 1938, prior to their setting out to find explorer Leichardt's grave. The photo of Isabella on the camel taken during the 1936 Ayers Rock expedition.[71]

from a distance and a view of the camp set up at the base. There is no footage of the Climb. In a strange twist Aborigines engaged to re-enact part of the story dressed in "war paint" were mistaken for hostile warriors by another group of prospectors, who reported spears were thrown at them.[72]

Cutlack's expedition, November 1939

Morley Cutlack led a number of speculative prospecting expeditions into the Petermann Ranges and surrounds in search of Lasseter's Reef, including the one "attacked" by natives dressed in war paint for the movie financed by Hugh Foy in 1936. The National Archives of Australia host a number of files about Cutlack[73] including some from the Commonwealth Investigations Branch dealing with concerns about his mining activities in central Australia.[74]

In 1939 he led an expedition into the Ayers Rock region to search once again for Lasseter's Reef. The group included author and travel writer Frank Clune, Victor Dumas (Cameleer) and mining expert Ernest Bails.[75] While waiting for Cutlack and NT and Commonwealth officials to arrive by plane Clune, Dumas and Bails climbed the Rock, leaving their names in the summit jar in November 1939.

Charles P. Mountford and Lauri Sheard, August 1940

It seems from his first visit to Ayers Rock in 1935 to act as secretary to the *Board of Enquiry into alleged ill-treatment of Aboriginals by Constable McKinnon and others*[76] anthropologist, Charles Mountford, developed a strong connection to the Rock, and a deep affection and appreciation of the Aboriginal people who lived around it. His book, *Ayers Rock: Its People, Their Beliefs and Their Art* (1965), and his published thesis, the masterpiece *Nomads of the Australian Desert* (1976) are unsurpassed in their depth of research, content, academic elaboration and explanation of Aboriginal religious beliefs at Ayers Rock. Mountford went to great lengths to obtain genuine stories of features at Ayers Rock and The Olgas from Aboriginals who had grown up in the region and had long term connections to these places. He did this over many decades. His popular book *Brown Men and Red Sand* (1948) recounts his visit to Ayers Rock with his young assistant Lauri E Sheard in August 1940, part of a wider expedition into the Central Desert region to conduct anthropological research. The volume of information and data collected over Mountford's lifetime in the field is staggering and leaves a legacy for future generations to investigate. The Mountford Sheard Collection at the South Australian Library runs over 120 shelf metres and contains Mountford's photographs, field notes, diaries, artworks and correspondence.[77] It has been inscribed on the UNESCO Australian Memory of the World Register.[78] Unfortunately much of the work is difficult to access for the general public as it has been deemed culturally sensitive by Aboriginal religious leaders.

Brown Men and Red Sand (1948) and Lauri Sheard's diary of the 1940 expedition[79] provide unique accounts of their climbing activities at Ayers Rock. Inscriptions at the summit cairn suggest they climbed together (see above), but both accounts indicate they took turns, climbing and investigating the summit with their Aboriginal guides who included Moanya, Matinya and Tjundaga. Of special interest to Mountford was the location of a waterhole "Uluru" located somewhere on the northern or eastern side of the summit plateau. Sheard's diary indicates Mountford climbed first on the 6th of August after investigating the caves lower down the slope. His group did not locate "Uluru" and hiked back down. Mountford recounts in *Brown Men and Red Sand*: *Although Matinya had not been to the top of Ayers Rock before, he felt quite sure, from the description the old man had given him, that he could find Uluru. But when he saw hundreds, if not thousands, of circular pot holes which had been weathered out of the trough like gutters, he became confused, and after much searching had to admit that he was beaten.*

Sheard climbed the next day, the 7th of August, and it seems the earlier trip freshened the guides' recollections and they had more success in locating "Uluru":

Proceeding to the climbing slope, this time the position reversed, I went up and Mountford stayed below taking photographs, using the colour

*cine camera. The climb was not terrifically steep, but it was jolly hard
work, as for a part of the way we went on hands and knees, but once over
the steeper pit… We were searching for "Uluru", a rock hole which the
natives say is never dry. Mountford could not find it yesterday, but I had
the advantage of the experience as the two natives with me had both been
with him yesterday and knew the direction he had taken. … The place is
immense and it is difficult to convey an idea of the size of the rock, but
it would be quite easy for members of a party to lose one another and not
meet again in hours of wandering. The only signs of life are several places,
where a few bloodwoods grow apparently out of the rock, and in a few
places there are clumps of spinifex. The descent was not terribly difficult
and although very steep in places we were able to walk most of the way
down. It was a hard morning's work but well worth the effort.*

The party left for Mt Olga on 9 August. On returning to Ayers Rock
on the 14th Mountford climbed again on what appears to have been
a very cold, windy winter's day determined to view the rockhole and
obtain stories about it from his guides. From *Brown Men and Red Sand*:

*When we were about half way up, and on the crest of the steepest saddle,
Matinya, who had been afraid for some time, showed signs of refusing to
go any further. He kept shouting that the wind would blow him over the
edge, and on occasions I was inclined to agree with him. Even then, as I
lay spreadeagled on the rock gathering breath for yet another few yards,
and looked up that ridge with nothing but open space either side, I almost
persuaded myself that the game was not worth the candle. But Matinya's
fear had put me on my mettle. There was no turning back, I had to keep
going. Now and again we found shallow depressions, just deep enough
to lie in and allow the furious icy blasts to sweep over the top of us. …
Once we were on the summit, however, and amid the deep gutters, we did
not feel the wind again.*

They went on to find "Uluru".

Mountford would return to Ayers Rock several more times and
climb again, most notably in 1950 with students from Knox Grammar.
In 1941 Lauri Sheard enlisted in the Royal Australian Air Force as an

airman. He was reported missing over New Guinea in 1942 following a flying battle and his death was confirmed in 1945. Sheard is remembered at the Port Moresby (Bomana) War Cemetery in New Guinea. The name of the *Mountford Sheard Collection* was chosen by Mountford to give due credit to Sheard's role in preserving his life's work.

Figure 14: Charles Mountford on a camel, 1935. Photograph taken during the 1935 'Commonwealth Board of Enquiry into the alleged ill-treatment of Aborigines near Ayers Rock'
(State Library of South Australia PRG 1218/34/73)

Lou Borgelt, Cliff Thompson, Tiger Tjalkalyirri and Mitjenkeri Mick, 30 June 1946

In June 1946 local Anangu men Tiger Tjalkalyirri and Mitjenkeri Mick acted as tourist guides and camel handlers to Adelaide motorcycle mechanic and amateur film maker Lou Borgelt and his friend Cliff Thompson on a month long journey to Ayers Rock and Mt Olga.[80] The party left from Tempe Downs on 19 June and would return to Hermannsburg via Angus Downs on 16 July. At Hermannsburg

Borgelt and Thompson continued on to Haasts Bluff where they assisted with the construction of a church for the Lutheran Mission.

Borgelt's movie footage of the group's ascent of Ayers Rock is a remarkable legacy of that trip. Restored by the Lutheran Archives, the grainy but stunning colour footage provides a visual diary of the trip. At Ayers Rock it shows the group on the climbing spur, joking about in puddles on the summit and horsing about at the small pile of stones that marks the summit. The footage includes shots of the glass jar at the summit and a close up of the list of names of the Foy Expedition. The respect, camaraderie and humour of the group shines through in this remarkable footage. Of the Climb, Borgelt noted in his diary:

> *Sunday 30/6/1946. … After short rest four of us set out and commence climb of Rock up west side. It looked precarious. I first strip myself of underclothing and then the job is one. We climbed like ants of flies up a wall and gradually we make the top. Took lot of films. Water holes everywhere, hundreds of them. We arrive at Cairn set up there. We open bottle and tin and note names of previous visitors, McKay expedition, CP Mountford, Foys Expedition, Constable McKinnon etc. We film all the different bits and enjoy it immensely. Then a quick run down. I take special sunset film on west wall, excellent it was. After a good tea I write diary by fire. Then a bath, Goodnight.*[81]

Borgelt returned to the Rock and climbed again in 1956, once again filming his adventures.[82]

Tiger Tjalkalyirri and Mitjenkeri Mick were the first Aboriginal men to have their names listed in the summit cairn. Tiger maintained a close connection with the Rock all his life and was active in the land rights movement in the 1970s and 1980s. He helped linguists and anthropologists to understand Anangu culture and was renowned for his knowledge of songs, dances and artefact manufacture.[83] He died in June 1985, prior to the hand back of Ayers Rock in October that year. Derek Roff (Head Ranger 1968-1985) recounts his entertaining tourists in camp grounds in the 1970s. It's a shame his free and open attitude to visitors climbing his Rock and his wicked sense of humour seem to have been lost by the current group of custodians.

Figure 15: Mitjenkeri Mick, Lou Borgelt and Tiger Tjalkalyirri horse about at the summit 30/6/1946. Still from video footage taken by Lou Borgelt 1946 (Courtesy Lutheran Archives)

Figure 16: Mitjenkeri Mick, Lou Borgelt and Tiger Tjalkalyirri at the summit 30/6/1946. Lou holding one of the log sheets from the coffee jar (Courtesy Lutheran Archives)

Tiger Tjalkalyirri from Derek's Oral history interview[84]

Interviewer: *Now, tell about the time when Tiger, that well-known entrepreneur, decided he would put on an individual dance performance, at a site at the Rock.*

Derek Roff: *… But he was a bit of a wag, was Tiger. I remember a number of occasions at the evening performances, people would be desperately trying to get Tiger to sing a portion of a religious, sacred, song. And Tiger would refuse — no he couldn't do that, and he couldn't do that. And he'd refuse and he'd refuse, and people would press him and press him, and he eventually, he would break. And he'd say: 'Alright, look, I've got to take you behind these bushes, and I'll sing a sacred song, but you can't tell anybody.' Everybody said [whispers]: 'Alright, alright, Tiger, no worries.' And he'd take them behind this bunch of bushes and he'd say, he's going to sing that sacred song about putcha. And everybody would sit there absolutely glued to the spot. And he'd start to sing and dance, and he'd start to sing: 'You putcha left leg out, you putcha left leg in.' [Laughs] Well, of course, they took it in very good part, because they knew they'd been had [laughs]. But he was a great character, a great character.*

Arthur Groom, Tiger Tjalkalyirri and Tamalji, 4 September 1947

One of the finest accounts of the climb is by Arthur Groom, a conservationist, author, and a remarkable bush walker who in many instances is recorded walking more than 100km in a day. In July 1947 he took a Qantas flight from near Brisbane to Alice Springs. He got a ride to Jay Creek and walked to Hermannsburg through the West MacDonnell Ranges, and then on to Areyonga and Tempe Downs. At Tempe Downs he met with aboriginal guides Tiger Tjalkalyirri, Tamalji and Njunowa, and on 25 August 1947 the group left for a visit to Ayers Rock and the Olgas. Their trek on camels was via the Levi Range, Kings Canyon and the eastern end of Lake Amadeus. The group climbed the Rock on 4 September. Groom returned to Henbury Station via Mt Conner, Angus Downs and Mr Ormerod.

Groom's tale of the Climb, especially his description of the reactions of his companions, wonderfully captures the joy, exhilaration and companionship of climbing. From Groom's *I Saw a Strange Land*:

Various writers have described Ayers Rock as difficult of ascent, when in reality it is a trained mountaineer's job on the east-south-east corner, a rough and steep scramble up at least two places on its southern side, and nothing else but a strenuous and spectacular uphill walk on its western side.

On the northern side, the sheer cliffs and hollowed base prevent any reasonable attempt at ascent.

We tackled the easy western route, called Tjinteritjinteringura (Willy Wagtail) by the natives. It is a bare rock ridge, not much steeper than a staircase, rising from a broad beginning to a narrowing ridge of sandstone, surfaced with the rough stucco pink common to the whole Rock. Tamalji soon gave his wild whoop and screaming laugh, and ran barefooted all over the place. His balance was amazing. Tiger grunted and groaned. 'Me getting old in legs,' he gasped; but he was determined. In desperation he turned to me. 'You feel all right? You not feel like properly old man?' But I was feeling splendid, and raced after Tamalji, who had taken the steep climb in his teeth. Tamalji was defying all known laws of extreme exertion by gulping several mouthfuls of ice-cold water from each round rock-hole. The man had the energy of a demon, and still found breath to laugh in wild abandon.

Njunowa gasped his way up the rock for perhaps a hundred yards then flopped on the Rock and rolled out flat on his back and lay spread-eagled for our return. Mount Olga's many domes rose above the sea of sandhills; but there were other mountains beyond, a strange, encircling concourse of rocky silhouettes and distant shapes of the far Petermanns to westward, and the Musgrave Ranges in a chain of peak after peak to southward. Some of them had never been trodden by white men. Two days' travel to westward was the lonely grave of prospector Henry Lasseter, who died of dysentery and starvation on 30 January 1931; buried crudely by friendly natives of the Petermann Ranges, reburied by old Bob Buck, Central Australian wanderer, who now lives at Doctor's

Stones in the eastern end of the James Range, south-west of Alice Springs. There was no trace of Lake Amadeus to the north. It was hidden in its salty hollow only six hundred feet above sea level.

The summit, crest, sides, ridges, ravines, shelves, and terraces of Ayers Rock are pitted with hundreds of the rounded rock-holes, capable of holding from a few to several thousand gallons of crystal-clear water from any light passing shower. The recent four inches of rain had filled every hole, until each pool overflowed to the next in a scintillating chain of flashing light.

Tamalji beat me to the top by a furlong. His time for the ascent was about thirty-five minutes. Tiger was still a dot turning the shoulder a quarter of a mile down. A small pile of broken sandstone has been placed on the summit, and the usual summit tin and bottle of names are there. I took out the pieces of parched and frayed paper. I have been on many a mountain summit, and seen many a cairn of stones and bottle filled with names; but none excited me more than those accounts of the past few who have travelled hundreds and in some instances, thousands of miles, to ascend Ayers Rock. In this lonely land it seemed to give the names written in ink and pencil definite reality and personal presence. Goodness knows where they all are now; but here are the names:

7/3/1931. W. McKinnon.

19/2/1932. W. McKinnon.

July, 1933. W. Fuller.

28/5/1936. H. N. Foy, Mrs Foy, Tom McFadden, Stan Tolhurst, Gus Schaller, Bill Morgan, Sydney Walker, Bob Buck, Denis Haycroft, Rupert Kathner, Kurt Johannsen, S. Mulladad.

Nov. 1939. V. Dumas, F. Clune, E. Bails.

7/8/1940. C. P. Mountford, J. E. Sheard.

14/8/1940. C. P. Mountford, J. E. Sheard.

30/6/1946. Lou A. Borgelt, Cliff Thompson, Tiger, Metingerie.

Tiger arrived and flopped straight into a pool to cool off. 'My legs get properly tight!' he grumbled, and thumped his cramped thighs. Tamalji run about too much like big euro!'

Groom's stunning photo of Tiger and Tamalji at the summit cairn, with Tamalji relaxing and Tiger's stoic dignified stance reflected in a waterhole, provides a solid rebuke to recent claims that the Anangu "never climb"[85] or don't climb the Rock.

Figure 17: Photo taken by Arthur Groom, 4 September 1947, showing climbing legend Tiger Tjalkalyirri and Tamalji (seated) at the summit[86] (National Library of Australia)

John Béchervaise, 23 September 1948

Groom was the last to travel any distance to visit the Rock by camel. In late 1948 a graded track following in part the trail blazed by Michael Terry and the Foy Party succeeded in reaching the Rock. Prior to its completion in November 1948, a small group led by Australian writer, photographer, artist, historian and explorer John Béchervaise, arrived at the Rock in September 1948 in a 4WD Chevrolet Blitz. His group left their names in the summit cairn on 23 September 1948. In the same trip Béchervaise also climbed Mt Olga on 5 September 1948, only the second reported to do so after Constable McKinnon in the 1930s.

1950-1958

Key figures in the development of tourism at Ayers Rock in this period include Len Tuit and Eddie Connellan. Tuit was instrumental in demonstrating the success of bus trips from Alice Springs and establishing accommodation and a permanent water source at the Rock while Connellan was successful in petitioning the Federal Government for a landing strip and providing air transport to Ayers Rock. This was constructed in early 1958.

The graded road to Ayers Rock, completed in November 1948, made the area much more accessible to conventional motor vehicles and the rate of visitation increased steadily as word of this natural wonder gained wider attention. Visitor numbers were still limited as access still required permission from the Department of the Interior. This changed in the late 1950s when the area was excised from the Aboriginal Reserve and formally declared a National Park in 1958 (Ayers Rock-Mount Olga National Park).

A few notable climbers in the period included:

- Knox Grammar School Excursion,[87] September 1950.
    - Included 22 students. Fastest students to climb to the summit did it in 25 minutes.
- Beryl Miles, Gordon Donkin,[88] Winter 1951.
    - Beryl was the second woman to record her name at the summit cairn, the first woman to be photographed on the summit.
- Cyril E Goode, Rex Ingamells and Bert Phillips, Christmas 1952.
    - Early Tourist road trip. The trio drove from Melbourne via Adelaide in a Rolls Royce Ute.
- Petticoat Safari, October 1957.[89]
    - Women's tour group covered by the *Australian Women's Weekly*. Oldest climber was grandmother Mrs Sarah Esnouf in her 80s.

Knox Grammar School Excursion, September 1950

The Knox Grammar School expedition to Ayers Rock in 1950 must surely count as one of the most ambitious school excursions in Australian educational history. A full account of the journey is provided in a series of articles by Dr Geoffrey Latham in the *Australasian Photo Review*[90] along with numerous contemporary newspaper reports.[91] Taking such a large group of people (six masters and 22 students) to remote parts of Central Australia at this period represented a considerable logistical challenge, requiring three years to plan. The group left Sydney by plane to Adelaide on 30 August and then by train (the Ghan) to the Finke River siding where they were met by Len Tuit, Ossie Andrews and Ron Dingwall with a small touring coach, blitz wagon and a three ton truck to cover the remaining 200 miles to the Rock. They arrived at Maggie Springs at 10pm on Tuesday, 5 September. They stayed for seven full days at the Rock, leaving for Alice Springs by vehicle on 13 September, arriving at Alice on Friday, 15 September. The group returned to Sydney by chartered plane late on Monday, 18 September. The expedition had a genuine scientific focus with student groups assigned tasks including geology, geography, biology and anthropology. The CSIRO provided scientific equipment for use in collecting samples. The group were led by T.W. Erskine and accompanied by anthropologist Charles Mountford. On Thursday, 7 September, the group tackled the "Great Climb". Latham recounts:

> *The Great climb occupied all of Thursday morning. On the western side there is the most gentle slope averaging about 45° but in places it probably exceeds 60° or so. Except for little flakes which may 'give' when trodden upon the rock surface is smooth, offering nothing upon which the hands can grip. Down each side of the climbing ridge the Rock slopes down precipitously to the ground. You would only slip once on that climb! This ridge finishes directly above the top of the ravine; away down below we could see the camp, barely visible at the end of the road as it snaked its way in and from the sand dunes.*

> *A short walk and a final climb completed the ascent to the cairn at the*

summit. Here, beneath a heap of stones, was a coffee jar serving to keep safe the names of previous climbers – practical and theoretical; some lists were obviously written by the one hand, and it is therefore open to doubt whether all those listed really climbed to the top. The names we found are given in the accompanying table (see above): some of them were very difficult to read and I apologise in advance for any errors.

There was little grass and a few shrubs growing at the top, while a group of boys claimed to have seen a wallaby. In some of the wind holes a little water remained. Three of us went out to the eastern end, but honour was hardly worth the effort, since the route is crossed by innumerable wind furrows. These measure up to fifteen feet deep, all running NW-SE. They are the result of differential weathering of the Rock's strata which are vertical to the ground. From the eastern end, Lake Amadeus could be seen as a thin white line extending along the horizon.

Two of the more energetic boys achieved the 1100ft climb in twenty five minutes; and another climbed twice in the one morning for a wager!

A colour movie of the expedition titled "Red Horizon", narrated by then prefect Edrich Chaffer, is available on Vimeo[92] and provides a wonderful overview of the expedition.[93] Edrich ends his narration about the expedition stating: *It set me on a career where I personally had learnt one of the most important lessons of all, and that was: how to think. And I found it invaluable in my career ever afterwards. And I can only thank Tom Erskine, William Bryden, our headmaster, for making the opportunity for young formative minds to be developed in such an excellent way.*

The scientific nature of the Knox Expedition has been mostly replaced with myth, superstition and political correctness that seem to dominate modern school visits to Ayers Rock. If the ban on climbing goes ahead they will soon not even be able to experience the joy, wonder and awe of climbing it. Without the spirit of science, adventure and discovery that characterised the Knox Expedition few students these days are likely to gain memories and experiences that will leave such a lasting, life-changing impression that the Knox boys experienced. A few Instagram moments of the "climb is closed" sign and being told "what to think" by muddle-headed tour leaders is unlikely to inspire the next generation of student visitors above mediocrity.

Latham and Erskine led another schoolboy expedition to Ayers Rock in 1952 with members from a number of high schools including Knox, Scots College and North Sydney Boys High School. Little has been written on this second school excursion.

Figure 18: Knox Grammar excursion. Ron Dingwall, Charles Mountford and Ian Tuit take a break at the summit, 7 September 1950. Dunlop volleys then the preferred choice for climbers (Photo courtesy Ron Dingwall)

Beryl Miles, Gordon Donkin, Winter 1951

Beryl Miles was a young English traveller intent on experiencing and writing a book about an adventure in Australia. In 1951 she was lucky to meet with photographer and adventurer Gordon Donkin and join him on a six month tour of Central and Northern Australia photographing Aboriginal art for the Anthropology Department of Sydney University. In Winter 1951 Beryl became the second woman to have her name recorded in the summit jar after Isabella Foy. By some accounts she was probably the first Western woman to climb, given uncertainty about the actual climbers in the Foy party. Her account of finding the jar at the summit is reminiscent of a modern geocaching hunt. From her 1954 book, *The Stars My Blanket*:[94]

> *I shall never forget the feeling of anticipation when, finally, we did reach the cairn and I bent down to lift off the first stone, then the next and the next, and then the horrible sinking feeling that followed as only more and more stones appeared underneath. Perhaps the bottle had been stolen.*
>
> *Then, suddenly, there was a glint of glass—two jars, both containing scraps of paper with names on—and a small flat cigarette-tin with the name 'McKinnon' scratched on the lid.*
>
> *As if to make things quite perfect, the very first paper I pulled out of the bottle read as follows:*
>
> *Wed. 4.9.47. Arthur Groom of Binnie Burie, Queensland and Tamali, Talkajeri (Tiger) of Hermannsburg Mission, leaving one native boy Erowa one third of the way up . . . via Lake Amadeus. Tomorrow we go to Mt. Olga. Signed: Arthur Groom (and a row of hieroglyphics from Tiger)'.*
>
> *It was a proud moment when I added my name to those in the bottle.*

Cyril E. Goode, Rex Ingamells and Bert Phillips, Christmas 1952

In January 1953 the *Centralian Advocate* records the flying visit of Cyril E Goode, Rex Ingamells and Bert Phillips to Alice Springs. Bert a school teacher and Cyril a well-regarded writer had left Melbourne some days previous driving *an ancient ... Rolls Royce buck board ute*, collecting poet Rex in Adelaide on the way through. The newspaper report is likely a few weeks old as the three were in the region around Christmas 1952 – not the best time to be travelling in Central Australia!

After re-supplying and re-fuelling in The Alice, the artistic but practically minded trio set off south to visit Ayers Rock and Mt Olga. Their epic journey was recounted in an essay by Cyril E.

Figure 19: Beryl Miles standing on Ayers Rock, 1951
(From *The Stars My Blanket*, 1954)

Goode, *Account of a Dash to Ayers Rock*, published in the somewhat obscure *Accademia Internazionale Leonardo da Vinci*, in 1972.[96]

This is probably the first Ayers Rock road trip on record, a journey that would be repeated by millions of time limited travellers seeking to explore the Rock, get a view, and climb the natural wonder at Australia's heart.

41

Goode's account of their time at the Rock is an enjoyable read and mentions climbs by Cyril above Maggie Springs and the three men climbing together to the summit via the conventional route. Goode writes:

> On reaching the top, the feeling of exultation made everything worthwhile . . . though the sense of tremendous space on all sides, in this rolling ocean of desert, cannot be properly described—at least not in an article of this length. The wild beauty of shimmering Mount Olga twenty miles away has to be seen to be believed. Other points of interest are: the Promised Lands of the mirages; the giant twenty- or thirty-foot corrugations running parallel across the bald surface; the shallow pools made by the recent storms and teeming with life: tadpoles, water-boatmen, mud shrimps and crimson dragonflies; the suicidal chasm leading down to Maggie Springs; the cairn of stones with the jar-full of famous names (Fellow FRANK CLUNE'S was the first I pulled out) . . . all these things repay the climber!

Rex Ingamells' published his own short account "Journey to the Rock" in *Walkabout Magazine*, May 1953. The article includes mention of another climb starting from Maggie Springs:

> Of the many entries written on scraps of paper in the two bottles at the cairn, I copied down one which risking the laws of copyright – confidently safe in the Brotherhood of the Rock – I reproduce "On the 16th September, 1952, Peter Magnus, 15, and Milton Osborne, 16, climbed Ayers Rock, starting from Maggie Springs. The climb took 1 hr. 30 mins. Both boys were members of the NSW Schoolboys Scientific Expedition. The boys came from Scots College and North Sydney Boys High School Respectively Signed Milton Osborne.

Ingamells' leaves a poignant reminder that closed minds are the greatest threat to future enjoyment at Ayers Rock: *It remains for individual discovery so long as the human mind retains its capacity for wonder.*

Petticoat Safari, October 1957

In the late 1950s, Trans Australia Airlines (TAA) established a "Women's Travel Advisory" section and in 1957 they organised a women's only trip to Ayers Rock as part of a strategy to promote tourism to central Australia. Billed as "Ann Travair's women's tour to the Alice and Beyond" the trip became known as the Petticoat Safari and was covered by a colour feature in the 23 October issue of the *Women's Weekly*.[97]

The group included Women from all over the country with ages ranging from 19 to 80.[98] They flew to Alice Springs and journeyed to the Rock by coach stopping at Mt Quinn overnight then onto the Rock the next day. The women spent their first day being guided around the northern side of the Rock by Bill Harney who been appointed as the Head Ranger in March of that year (the National Park to be formally gazetted early in 1958). They tackled the climb the following day and then the south side of the Rock with Bill the next. Then the group went by bus to the Olgas the following day, the road having been substantially improved, and back to the Rock for their final night. They left the Rock on the 10th of October, stopping at Mt Quinn overnight again and arrived back at Alice Springs on Friday the 11th. In 1958 similar bus trips were able to reach the Rock in a day with the road being gradually improved.

Of the Climb the *Women's Weekly* reported:

First girl to climb the mountainside was Evelyn Camm, former dressmaker, now a Melbourne tram conductress, who said balancing on trams could have helped her balance on the Rock.

And first "safari" girls to write names on paper, enclosing them in jam-tins or bottles at the stone cairn, were Colleen Lewin, of Tasmania, and Victorians Esma Davis, Jean Mason, and Nancie Reed. Lying around, sucking mandarins and admiring the view and the soaring eagles above, we were all feeling pretty smug about our achievement, when over the rise came grandmother Mrs. Sarah Esnouf, of Melbourne, helped by Peter Watts and Ian Lovegrove. Mrs. Esnouf, one of the oldest in the party, joined us, crying, "How exhilarating. Have you ever seen anything so wonderful? And look at the beautiful wildflower I picked on the way.

Edna Bradley (Saunders) was a member of the original Petticoat Safari. Her memoir, *A Rock to Remember*, a wonderful account of that trip and of early tourism at the Rock. Following her 1957 trip Edna would return the following year to work for Len Tuit on the Ayers Rock tour and other tours out of Alice Springs. The day after climbing to the summit with the rest of the Petticoat Safari group in 1957, Edna woke early and along with her friend Yvonne, and led by tour guides and drivers Peter, Ian and Dave climbed the Rock once more; this time to explore the deep valley above the Mutitjulu waterhole:

At the top, instead of turning left and going across the ridges to the cairn, we went straight ahead walking on the tops of the ridges. So far so good, but then the slope on the other side was much steeper and we all resorted to sliding on our backsides hanging onto whatever hand and footholds we could find. …The last few feet before we dropped onto the shady ravine became very smooth and almost vertical, but somehow we managed to scramble down. … There were bits of shale to hold onto which stopped my downward slide and finally my feet touched some ground. With a sigh of relief. I thought, at least I got this far safely. The boys were already heading towards the end of the ravine. I was surprised at how big it was, it went right into the Rock and was about 7 meters wide with the sides towering over us.

This is the first account of a tourist group climbing down into the gorge above the Mutitjulu water hole. Access to this area has been banned for some time. The gorge is spectacular and it would make great *Via ferrata* route, perhaps starting from Mutitjulu Waterhole. On their safe return after having to climb over each other on the way out Edna writes: *The realisation of where we had been made us feel special, but Ayers Rock robs you of ego. Looking up at the size of the Rock, I felt as small as an ant.*

History of Climbing Part 3: 1958-1985

Formation of the Ayers Rock-Mt Olga National Park in 1958 opened up the area to general tourism with visitor numbers

Figure 20: Petticoat Safari at the top of Ayers Rock
(Picture courtesy Edna Bradley (Saunders), 4th from the right.)

increasing each year up to the early 2000s with the Climb always being the main reason why tourists visited. The end of growth coincided with Parks Australia adopting a more stringent management plan in 2001 devised by the Board of Management that emphasised *Tjukurpa above all else.*[99] The effect on tourist numbers was quite immediate and after nearly 20 years they have not recovered to previous highs.

The new Park was managed by the Northern Territory Reserves Board which appointed Bill Harney as the first Ranger in March 1957 prior to the Park being formally declared[100] in early 1958. Over the first few years the Park ran a seven month long tourist season (April-October) and the position of Ranger was part time. Harney retired on 15 October 1961 to Queensland, passing away on the last day of 1962 at his home at Mooloolaba. Harney was a well-regarded bushman and story teller and his interpretations of Aboriginal myths and legends picked up from two local Aboriginal men (Kadakadeka and Imalung) formed the basis of Aboriginal myths and legends about Ayers Rock in early tourist guidebooks. Harney was replaced briefly by Jack Street who acted as curator until 3 October 1963 when Bob

Gregory was appointed and the position became a full time one. Gregory served until October 1968 when the job of head Ranger was taken up by ex-Kenyan Policeman Derek Roff. Above anyone else in this period and after, Roff played a major role in successfully juggling the needs of tourists, the claims of Traditional Custodians and an increasingly hostile political battle between the Territory and Federal Government over control of the Rock. Roff's exemplary tenure lasted until the Federal Government took management of the Park away from the Conservation Commission of the Northern Territory following the handover to Traditional Custodians in 1985. A request to Parks Australia for details of head Rangers who served after Roff resulted in the following response from Park Manager Mike Misso: *We don't have a definitive list that I am aware of and many of the early records are unlikely to (now) be in our HR system.*[101] It seems no one remembers face-less bureaucrats.

The name of the Park was changed on 24 May 1977 to Uluru (Ayers Rock-Mt Olga) National Park, and again in 1993 to Uluru-Kata Tjuta National Park.

Cairn and log books

On 15 May 1958[102] the small pile of stones that William Gosse erected in 1873 was replaced by a larger formal survey cairn, complete with post and vanes constructed by the Australian Division of National Mapping during their Australia Geodetic survey.[103] In 1970 this was in turn replaced by a formal bronze directional plaque and pedestal made by Melbourne firm Mechanised Methods that included a bronze shelf to stow a log book that recorded climbers' names. The plate featured the Australian Coat of Arms, a map of Australia and directions and distances to nearby ranges and other natural features. The bronze work was encapsulated into a four foot high stone pedestal. This remains in place today, though it is now in desperate need of restoration and under threat of removal after the ban is in place. The island of Tasmania on the map was lost in the early 1970s, and the rest of Australia was removed in the early

1980s. The Australian coat of arms survived into the 21st century but photographic evidence indicates it was lost in the early 2000s. The stone pedestal surrounding the bronze work was constructed by Northern Territory Parks staff Derek Roff, Ian Cawood, George Page-Sharpe, Darrel Toon and Ian Dawkins, with oversight from National Mapping's Surveyor Bill Johnson to ensure it went in the right place. The stone used for the surrounds was sourced from Mt Conner. Johnson recounts, in a letter to the National Mapping Director, that the Bronze plaque contained a spelling error not picked up until after the plaque was erected:

> *A point just noticed whilst drafting these letters and collating the photographs is the spelling of "Territories" in the circular legend "Erected in collaboration with Northern Territories Reserves Board" – it should be "Territory". It was on the diagram sent to Mechanized Methods, despite the repeated checking, and it passed unnoticed apparently even by the Board's personnel who all seemed to be pleased with this beautiful piece of metallurgical art.*[104]

**Figure 21: The Summit Cairn, end of 1970
(Photo taken by Laurie McLean, courtesy xnatmap[105])**

Formal log books recording the names of climbers replaced the assorted collection of jars and tins lodged at the cairn in

**Figure 22: Bronze Directional Plate at the top of the Rock
December 1970, waiting installation
(Used with permission National Archives of Australia BC 8241845)**

**Figure 23: Left – Summit pedestal under construction, December
1970. L-R George Page-Sharpe, Derek Roff (Head Ranger), Ian
Cawood and Darrel Toon (thanks to Roberta Roff for the names).
Right – Completed memorial with logbook shelf
(Photos used with permission of National Archives of Australia
BC 8241845)**

1966. These were used into the mid-1980s until Parks Australia, apparently without consultation with the visiting public, stopped maintaining them. Shamefully in the 2000s the log book shelf in the pedestal was clumsily sealed over preventing access – an act of vandalism concealing cultural history from public view. The future of the pedestal is now also under threat as it faces removal from the Rock, along with the chain and the five memorial plates to people who died on the Rock.

The 171 logbooks for the period between 20 May 1966- 24 May 1986 are currently stored at the Northern Territory Archive Service repository in Alice Springs. Until the pedestal was completed in 1970, log books were housed in a tin container on the stone survey cairn. Their contents make for fascinating reading, containing inspirational messages from millions of Australian and international visitors who have climbed the Rock. The whereabouts of the jars left by McKinnon with the slips of paper signed by the first climbers is unknown.

Signs

Signage at the base of the Rock did not change much up to the end of 1985. The small sign, essentially warning climbers they are responsible for their own risk, was replaced by a slightly larger sign highlighting some of the dangers. The "do not climb" and "we never climb" message would first appear in 1991 and signage and fencing at the base has become increasingly prescriptive and draconian since then.

The Chain

Concern about tourist safety following two deaths in 1962 and 1963 resulted in the installation of a chain up the climbing spur, and a white line being painted to guide walkers safely across the top of the Rock to the summit. This was installed in two sections in 1964[106] and extended and joined to form a continuous chain in 1976[107] and extended a little in the early 1980s. There are 136 posts in the main

NOTICE

THE PUBLIC IS HEREBY NOTIFIED THAT THE CLIMBING OF THIS ROCK IS A DIFFICULT AND DANGEROUS FEAT AND THAT THIS BOARD ACCEPTS NO RESPONSIBILITY FOR INJURY OR LOSS OF LIFE, TO PERSONS ENGAGED IN CLIMBING THE ROCK

RESCUE GEAR IS AVAILABLE AT THE CURATORS COTTAGE·
BY ORDER OF THE BOARD · A·PROSE·CHAIRMAN

NATIONAL PARKS AND GARDENS ORDINANCE
PUBLIC NOTICE
THIS AREA IS ONE OF THE RESERVES DEDICATED TO PRESERVING OUR SCENIC, SCIENTIFIC AND HISTORIC HERITAGE FOR PUBLIC ENJOYMENT

Figure 24: Sign at the base of the Climb, 4 December 1969. The small notice below reads: *National Parks and Gardens Ordinance. Public Notice. This area is one of the reserves dedicated to preserving our scenic, scientific and historic heritage for public enjoyment.* **Not for much longer!**
(Photo used with permission of National Archives of Australia BC 8241845)

section and another two in a short step up to the summit plateau beyond. The fate of the chain after the ban is not known. At the time of writing we understand the Board intend to remove the chain immediately after the Climb is banned.

Memorial Plaques

In this period five plaques were installed near the base of the Climb in memory of deaths at the Rock. From oldest the plaques read:

> In memory of Brian Streiff
> Who lost his life in a fall
> While climbing Ayers Rock

On 26th May 1962
Brian came from
Malvern, Victoria
And was visiting the rock
With a party of schoolboys
From Carey Baptist Grammar School.

In memory of Marcia Buniston
(Aged 25 years)
Of Yorkshire, England
Who lost her life
While climbing Ayers Rock
On 22nd December 1963.

In memory of Leslie Arthur Thwaites
Of Newcastle, NSW
Who died on top of this rock
On 15th June 1972
Aged 63 years
The climbing of Ayers Rock
was one of his lifelong ambitions.

In loving memory of Ernest Francis George
Died on Ayers Rock
10-8-77
Aged 59 years
Remembered by wife and family
And friends at
Liverpool Railway Station, NSW
"Rest in peace".
Plaque E Evans phone 046-25-6994.

In memory Brian Joseph Miller
Who died from a fall
While climbing Ayers Rock
On May 18th 1978
Aged 25
Sad loss to Jack and Jean Miller
Brothers Keith and Peter.

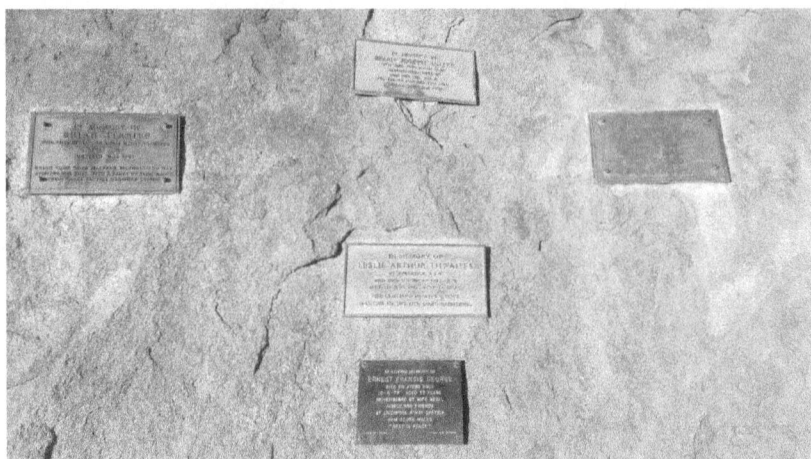

Figure 25: Memorial Plaques to climbers who died 1962-1977, at base of climb (Photo Zoe Hendrickx)

The 1967 Northern Territory Reserves Board tourist booklet mentions a memorial sign to Bill Harney located at the base of the Climb.[108] The sign was probably removed sometime in the 1970s. Parks Australia found an anonymous, undated photo of the sign, but could shed no light on its current location. A fitting replacement would read "Tiger Tjalkalyirri Lookout."

Figure 26: Sign erected to the memory of Bill Harney, first resident Ranger of the Rock (Photographer unknown. From UKNP photo library, used with permission)

Accommodation and access

Up until 1984 tourist accommodation was located on the east side of the Rock, in the vicinity of the current township of Mutitjulu. The various tourist buildings arose in an ad hoc way as demand increased and more facilities were required. A map in the 1967 guide book shows the distribution of the main hotels, camping areas, roads and walking tracks. The area was only about 2.5 km from the Rock. The main road from Curtin Springs came into the developed area from the south end.

The airstrip built by Peter Severin for Eddie Connellan was located close to the Rock on the North side, providing a spectacular viewing location for landing and departing aircraft. The first plane to officially use the strip was a Cessna 180 piloted by Connellen on 20 April 1958. This area has been remediated and a new run way capable of landing large passenger jets was built north of Yulara and officially opened by Malcolm Fraser on 6 June 1982 (Ayers Rock Airport or Connellan Airport).

Public campgrounds inside the park were closed in mid-1983 and the motels closed in late 1984 as tourist facilities were moved to Yulara Village outside the Park. Road access was reconstructed to provide for the new layout.

Tourist Information

Among the first tourist brochures for Ayers Rock and Mt Olga were those produced in 1961 by Charles Mountford and Ainslie Roberts.[109] These featured important Aboriginal myths and legends, and maps showing the location of the main features and advice on how and where to climb. Of the Ayers Rock climb the guide stated:

> But, although one can motor around the base of the Rock on a well graded road, walking is by far the best way of enjoying the majesty and grandeur of the surroundings. To climb the Rock is a much more strenuous task, but well within the ability of anyone in good health. Many hundreds, possibly thousands of people have made this climb and enjoyed the view from the summit. Although Gosse and Kamran

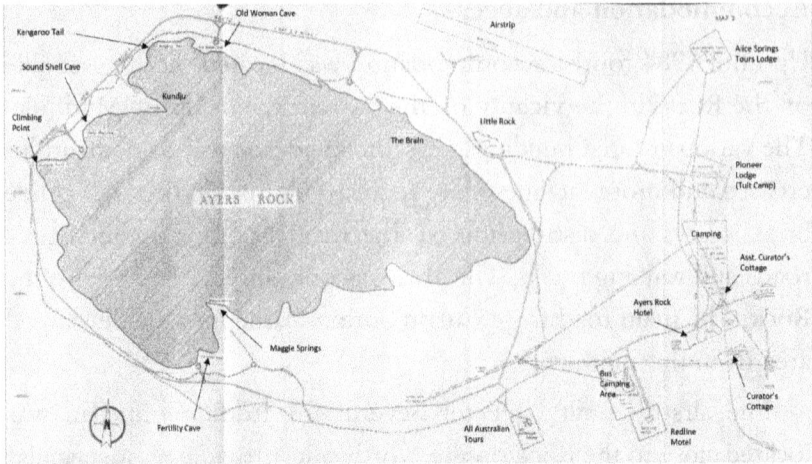

Figure 27: Map of Ayers Rock, showing location of main tourist facilities in 1967[110]
(Used with permission of National Archives of Australia BC8241845)

reached the summit of Ayers Rock by climbing up the Mutijilda gorge above Maggie Springs, the safer and much less strenuous task is along the slope on the western face.

The 1961 guide for Mt Olga included detailed advice on how to climb to the summit of Mt Olga, an activity sadly banned in 1986:[111] *Mt Olga is a challenge to all rock climbers. Although its ascent is much more strenuous and difficult than that of Ayers Rock, the effort involved is fully re-paid by the breathtaking panorama from the summit....* Mt Olga and some of the other domes would make for spectacular *Via ferrata* routes.

A booklet printed in 1967 by the Northern Territory Reserves Board may be viewed at the Australian Archives. It was included in the file documenting construction of the summit pedestal and directional plaque (Ayers Rock Geodetic Station).[112] This 40 page booklet was one of series of guides produced for the NT Reserves Board, the first was on the historic Alice Springs Telegraph site. The Climb is only briefly mentioned: ... *the only safe way of reaching the summit is via "The Climb" at the western extremity although at least two people are known to have made the ascent from Maggie's Springs.*

The longest serving head ranger, Derek Roff (1968-1985), pro-

54

duced a wonderful book of photography in 1979.[113] Of the Climb it stated:

> *The oldest man to climb the Rock was eighty-nine and the oldest lady eighty-two. One young man with a broken leg in a cast went up and down the Rock on his lower anatomy. Covering a distance of 1.6 kilometres and rising 348 metres above the surrounding plains, the Climb at Ayers Rock throws a challenge to park visitors. Thousands of people make the effort to climb the Rock and a large number are successful. Over the years there have, unfortunately, been accidents caused by falls and a number of persons have suffered heart attacks so intending climbers should ensure they fully understand the risks if they are not physically fit. They should also stay on the authorised climbing route where they will be reasonably safe.*[114]

An official guide to the park from May 1981 highlights the dramatic changes in attitudes and visitor access that have occurred since 1985. This has considerably diminished visitor opportunities and overall enjoyment and experience of the Park. The gold standard for any future management plan should be a return to the access that was available at this time. The 1981 guide prominently features The Climb on the cover. The notes indicate average climb times of about one-and-a-half hours to the summit with a record of 25 minutes return or 12 minutes up to the summit. It mentions 12 deaths on the Rock; four falls and eight heart attacks. It estimates 75% of visitors climb and provides practical advice on what to wear and how to get there. The section ends with sound advice: *If you made it to the top, congratulations, if you don't you probably have a very good reason. The saying holds true; "T'is a wise man that knows his own limitations."* The accompanying maps shows how access has been severely curtailed with significant attractions now quarantined from public view. The base walk is shown sticking close the base. The northern section today is well beyond the old circuit road. At the Olgas the walks through these majestic Domes are now constrained to short sections at Walpa Gorge and the Valley of the Winds Walk.

In the 1980s visitors could enjoy a walk through the Domes and also visit the Kata Tjuta Lookout described by Mountford in his 1961 guidebook as having an outlook *which has no equal in Australia*. These have been shut down for ideological and political reasons.

Prince Hiro climbs the Rock

One of the more notable climbers in this era was Crown Prince Naruhito (Hiro) of Japan. As part of a three week tour to practise English and give him a taste of Australia the 14-year-old prince and soon to be future Emperor of Japan climbed Ayers Rock on 20 August 1974. His climb, along with a culture of climbing mountains and mountain worship, is reason for the popularity of the Climb with Japanese tourists. To mark his visit the Japanese imperial family donated a wonderful 2x2m painting of Ayers Rock by artist Hideo Nishiyama.[115] We found it hanging in storage bay 150 at the National Library in Canberra. A red rock sitting in a mottled, swirling sea of purple, blue sand capturing the changing colours of the Rock at sunset. Oddly hanging adjacent to this remarkable, neglected painting and looking over it was a portrait of Lindy Chamberlain by artist Neville Dawson.

Aboriginal attitudes to climbing.

The attitude of traditional custodians to the Climb over this period is best summed by the Principal Owner of the Rock, Paddy Uluru. In an interview with Alice Springs journalist Erwin Chlanda,[116] Uluru indicated the act of climbing was of no cultural interest:

I had the opportunity of speaking with Paddy Uluru early in my work in Central Australia (I arrived in December 1974). Mr Uluru was the undisputed custodian of The Rock at that time. We spoke face to face at the base of monolith, and he was happy to be photographed with the Rock in the background. Mr Uluru told me if tourists are stupid enough to climb the Rock, they're welcome to it. For him there was nothing of practical value up there such as water, game nor edible plants. He made it clear that knowledge of certain elements of the Rock's dreaming must remain secret, to be

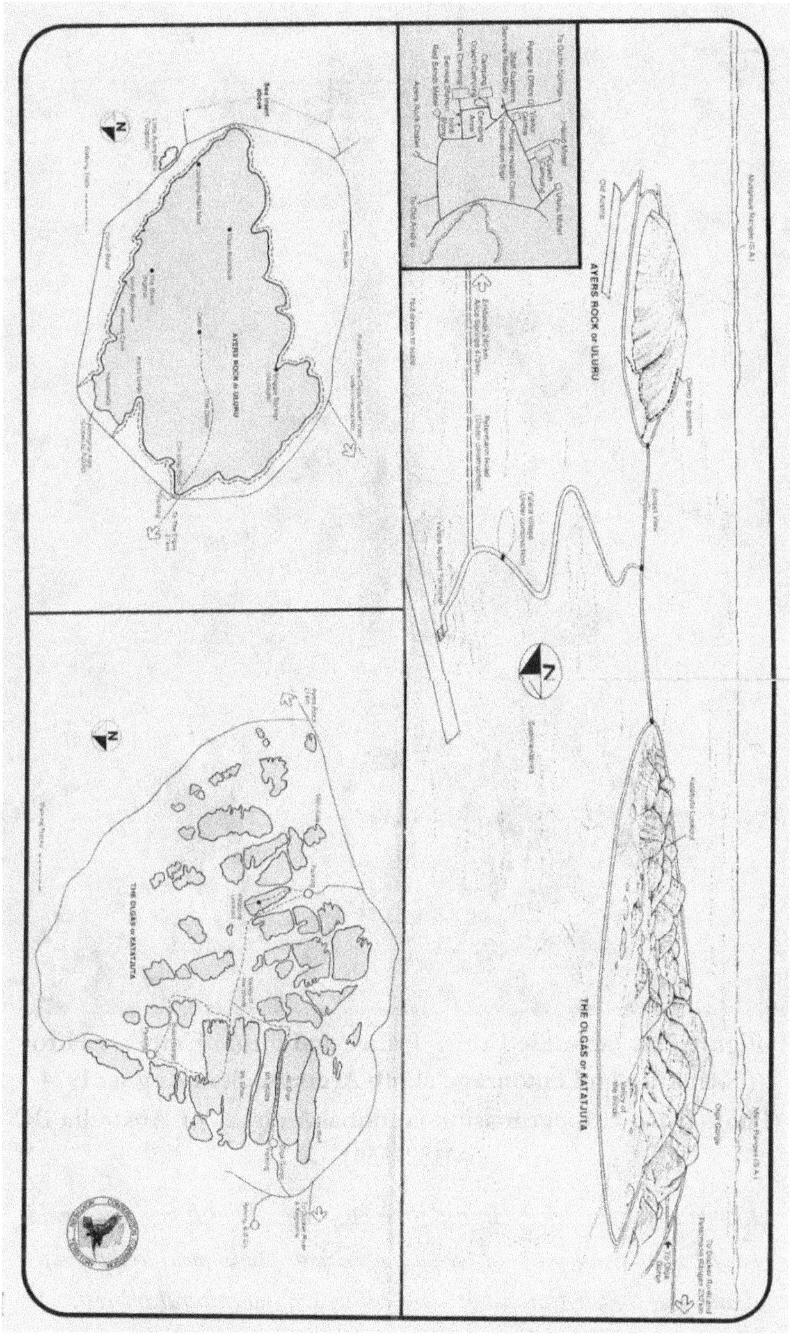

Figure 28: Uluru Guide 1981 map of the park showing much greater access compared to present (Conservation Commission of the Northern Territory)

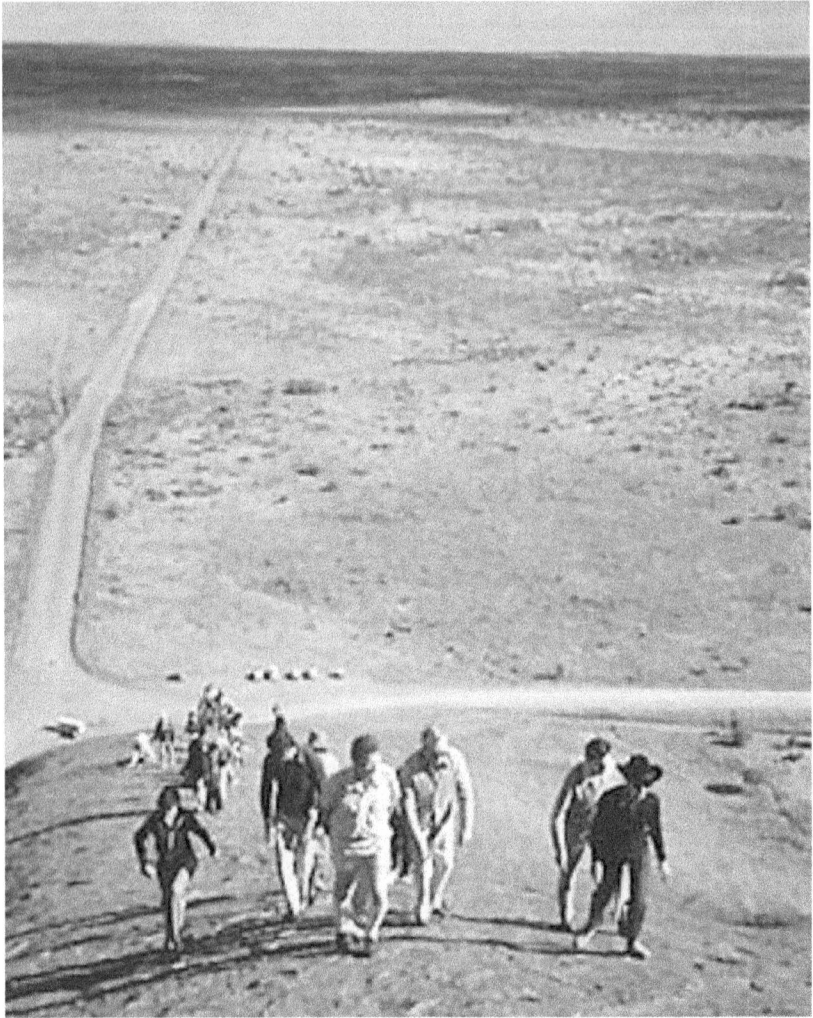

Figure 29a: Japanese Crown Prince and future Emperor Hiro (Naruhito) and entourage climb Ayers Rock in August 1974 (Photo used with permission National Archives of Australia BC 11827554)

known only by a strictly defined circle of people. That knowledge would be passed on to outsiders at the pain of serious punishment and perhaps death. But the physical act of climbing was of no cultural interest, Mr Uluru told me.

These views were confirmed by Head Ranger Derek Roff in an interview with the Northern Territory Oral History Unit.[117]

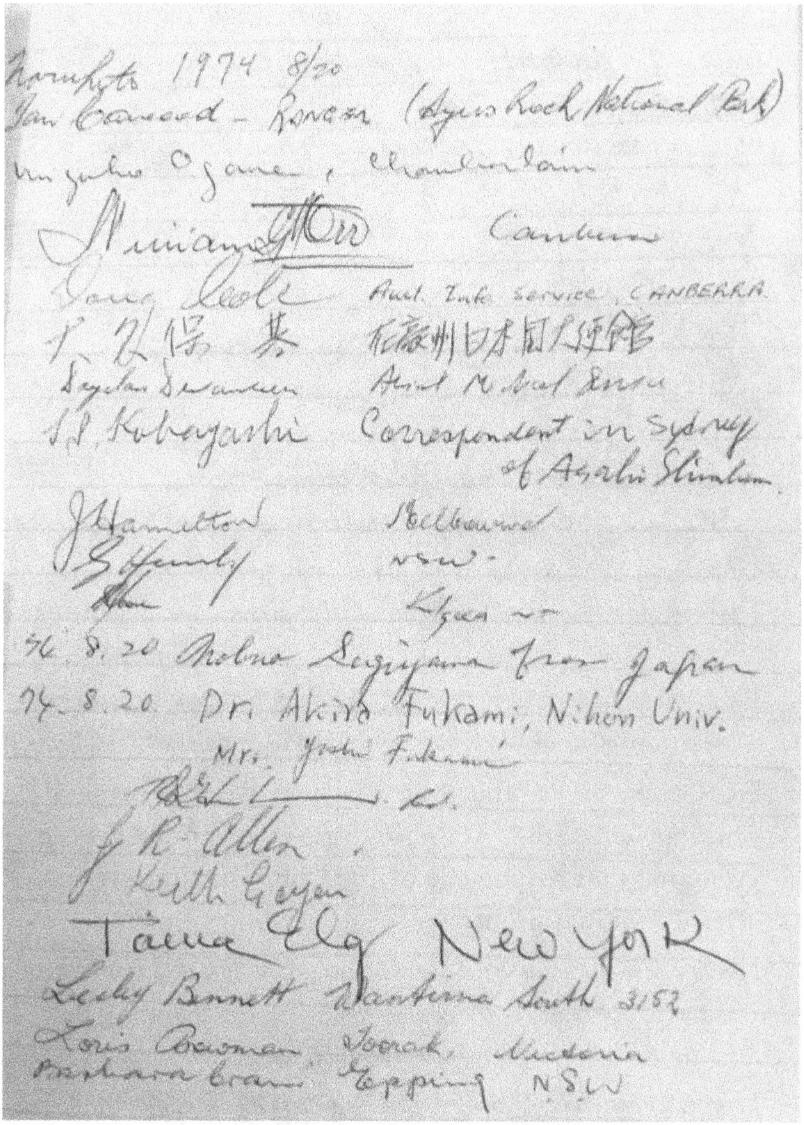
Naruhito 1974 8/20
Ian Cawood - Ranger (Ayers Rock National Park)
Canberra
Aust. Info Service, CANBERRA
S.J. Kobayashi Correspondent in Sydney of Asahi Shimbun
J Hamilton Melbourne
NSW
74. 8. 20. Dr. Akira Fukami, Nihon Univ.
J R Allen
Keith Goyen
New York
Lesley Bennett Wantirna South 3152
Toorak, Victoria
Epping NSW

Figure 29b: Naruhito's modest entry from the logbook at the Cairn on 20 August 1974, Ian Cawood (Ranger) immediately below (Logbook #38, used with permission NT Archives Service)

Roff was asked:

Did the Aboriginal people ever talk to you about either regulating the use of, or even closing the Climb?

Derek's response: *No. No, they never did. That never*

came up. There's often a statement I've read recently, and what-have-you, that the Aboriginal people never climbed it. Now, that, to me, doesn't quite ring true, (a) because there a number of sites on the top of course, that have got stories and names and what-have-you. Paddy Uluru used to tell me about climbing the Rock. It seemed to me that it was mainly the senior, traditional people who climbed, rather than everybody. But there was no doubt about it, that ceremonies were carried out in certain areas up there, that people did climb it. I'm just trying to think of the name of the Aboriginal people who went up with Mountford — Lively — Lively Pakalinga, Nipper's brother, older brother. He climbed it with Mountford, and explained some of the stories up there and what-have-you. So, I must say, certainly it was climbed — not maybe by everybody, but certainly the traditional people. And they never, you know, Paddy Uluru never mentioned the possibility of that to me. And I think, if he had had a concern about it, he would have, as he did with the cave. But I don't know, I must say, there it is. It's a thing that's come up now, by a different group of people. Well, fair enough. They've got their interpretation.

So, Derek, all of the time that you were living and working with the Aboriginal people, nobody ever intimated that maybe use of the Climb by tourists was offensive or inappropriate?

Not to me. No, they didn't. I must say, in actual fact, that was where the name for the tourists came up — 'minga' — it was watching them climb the Rock. And it was more a sense of fun than anything else.

I remember old Paddy Uluru being at the bottom of the Climb one day, with him, and we were just talking about this and that. Alongside the Climb is a sacred path. And we were just talking about this and that, and one of the American blue rinse set came up to him (and rather effusive, as they used to get) and she said: 'Oh, you going to the Climb the Rock?'

And he said: 'No, missus. If I go up there, I might fall on me f-ing arse!' [Laughs] Well, that brought the conversation to a fairly rapid close. [Laughter] Paddy was a man of many words.

On 13 February 1975, ABC's *This Day Tonight* broadcast a report by Grahame Wilson about the environmental impact of tourists at Ayers Rock and changes being proposed for accommodation by the Northern Territory Reserves Board. The daggy hotels and unsightly camp grounds would be moved outside the Park to a new resort complex 12km away. The report also covered the construction of a new fence to keep tourists out of the men's initiation cave (Warayuki) and adjacent Ngaltawata area (commonly referred to at the time as the Kangaroo Tail).

Wilson interviewed Yankunytjatjara elder and brother of Paddy Uluru, Toby Naninga. Transcript of that 1 minute 48 second section appears below. While Toby was happy to see the end of tourists wandering through Warayuki he had no objections to tourists visiting other areas of ritual significance, the Climb included.

Footage shows men on thongs digging holes for fence posts.

Grahame Wilson (GW): Fences are not a usual feature of a wilderness area of a National Park, but this fence is one sign of the new attitude to the real priorities around the landmark. The Yankunytjatjara people were remarkably tolerant of tourist intrusion into traditional areas. But they made one definite statement: no tourists at Warayuki the initiation cave or the sacred site Ngaltawata nearby.

Cut to Grahame and Toby Naninga standing on road next to fence, Ayers Rock in background, close up on Toby.

GW: Do you tell me, what is this place here?

Toby Naninga (TN): *Points to features.* Ngaltawata and Warayuki, Warayuki. Put a fence and gate up. No more tourists through Warayuki. Finish.

GW: You don't want tourists to go in there?

TN: Yeah, Yeah.

Cut back to Grahame and Toby looking at the Rock

GW: Toby Naninga is the brother of the Traditional Keeper of the Rock. This week he and two other members of the family will join the staff of the Northern Territory Reserves Board as park

Figure 30: Yankunytjatjara elder Toby Naninga and *This Day Tonight* Reporter Grahame Wilson discuss tourist access at Ayers Rock (Still from TDT footage broadcast by ABC, 13 February 1975)

rangers watching over their people's territory. Tourists will no longer visit the initiation cave still of great importance to the living Yankunytjatjara culture. But there's no objection to tourists visiting other areas of ritual significance, perhaps because vandalism has never been a serious problem here.

Close up on Toby.

GW: This is a very important place to you?

TN: Yeah.

GW: Do you mind tourists going anywhere else?

TN: Yeah anywhere, anywhere alright, you know...road, on cairn(?), tourist guide, anywhere.

GW: Anywhere else is alright?

TN: Yeah.

GW: But not here?

TN: No.

After Paddy Uluru died in 1979, ownership of the Rock was passed on to a wider group. One of these was Tony Tjamiwa. Tjamiwa was a respected Aboriginal Elder at Uluru. He was a board member of the National Park. When he died in 2001 the Climb was closed for 11 days. Tony's quotes feature prominently in Parks Australia Management plans and brochures and indicate climbing is not a "proper tradition". On the day of the handover (26 October 1985) Tjamiwa produced a map of the main sites at Uluru.[118] Tony's map shows a number of sites along the base of the Rock, but all it shows for the summit is a blue line that ends in a box labelled "Minga Line". The term "Minga" used by the Anangu to compare climbing tourists with ants. It seems the summit belongs to the Minga!

In the lead up to the handover a joint statement about aboriginal views of tourism at Ayers rock was made by the Central Land Council and Pintjantjatjara Council in November 1983. Read into Hansard by Minister Clive Holding,[119] it stated:

The CLC and the Pitjantjatjara Council are extremely concerned that the enlightened gesture of the Commonwealth Government in granting Aboriginal people title to Uluru National Park has already been distorted by the NT Chief Minister Mr Everingham for perceived political advantage.

Before the facts are further muddied in the NT election campaign it is essential that the position of the traditional Aboriginal owners is clearly stated.

- *The Aboriginal people have always recognised the legitimate tourist interest in the national park.*
- *They have always supported the concepts of leasing back the park to the Commonwealth.*
- *They have consistently asserted that the park will always be available for the benefit of all Australians.*
- *They have always supported a joint management scheme in which Aboriginal, conservationist and tourist interests would be represented.*

- *They have no intention of unreasonably limiting access to Uluru National Park.*

- *Basically for the visiting tourist it will be business as usual.*

- *Any rare and limited restrictions necessary for ceremonial purposes are likely to be confined to those sites already registered as sacred by the NT Government's own Sacred Sites Authority (and already subject to restrictions).*

- *Such ceremonies should be respected as a vital part of traditional Aboriginal life.*

- *The Aboriginal traditional owners believe that Aboriginal ownership and involvement in Uluru substantially enhances the commercial tourist potential of the park.*

- *The Yulara project will not be affected by Aboriginal ownership of Uluru. The Aboriginal people have expressed no interest in seeking to operate motels within the national park.*

- *Indeed, Aboriginal traditional owners welcome the Yulara project in that it locates tourists away from their local Mutitjulu community and thereby reduces the impact of thousands of tourists a year on their way of life.*

It follows that the granting of title to the Aboriginal traditional owners will not jeopardise investment in the Yulara operation.

The Hawke initiative is an excellent measure which recognises the long-standing spiritual attachment of the Aboriginal people to this area whilst preserving the interests of tourists and conservationists in the park.

With the introduction of the "we never climb" message in the 1991 management plan, it only took six years (1985-1991) for these promises to be broken. It seems there is nothing to indicate the route of the Climb was especially sacred to Traditional Custodians prior to 1991.

The period ends with the handover of the Park to Traditional Custodians on 26 October 1985 resulting in dramatic changes in Park Management that may ultimately lead to The Climb, this wonderful challenging engagement with the natural world, being banned altogether.

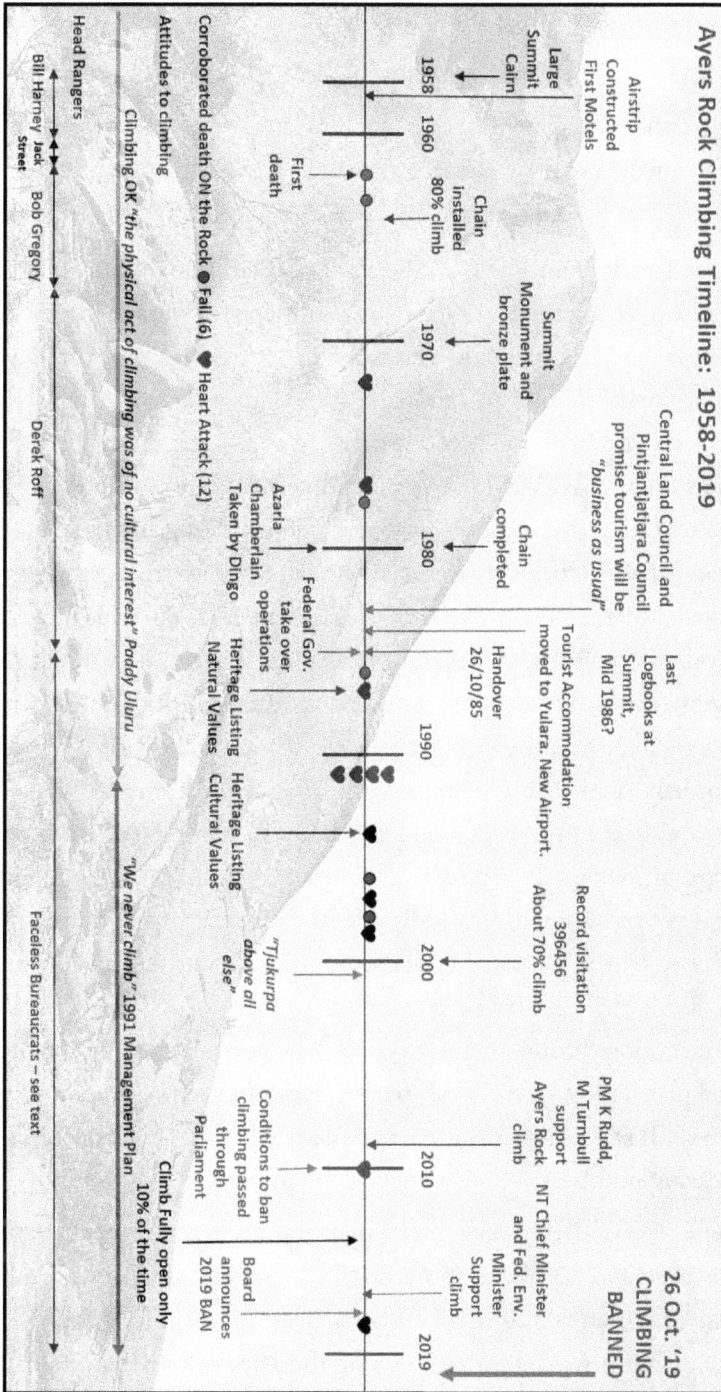

Figure 31: Summary of climbing history and related events 1958-2019

History of Climbing Part 4: 1985-2018.

Decline – A National no go zone emerges

This period saw many changes to the Climb, Park infrastructure and management, most of them detrimental to visitor enjoyment of the Park. The modern day experience is a shadow of what it used to be.

The formal handover of the Park to Traditional Custodians occurred on 26 October 1985. It was immediately leased back to the Director the National Parks and Wildlife Service (NPWS) for 99 years for continued use as a National Park. Management responsibilities were removed from the Northern Territory Government and handed over to a new Board of Management dominated by Traditional Custodians with day-to-day business and implementation being taken over by the NPWS (currently Parks Australia). A new management plan produced by the new board was released late in 1986. This included a ban on climbing the domes at Mt Olga but for the Ayers Rock climb no changes were proposed compared to earlier plans. The 1991 management plan was the first to indicate dissatisfaction of Traditional Custodians with the Climb and the first to promote a "We never climb" message and discourage visitors from climbing. It prompted a change of signage at the base. The reason behind the change in attitude was not outlined. Ironically a 1991 tourist information pamphlet produced by the National Park did not include any negative information about the Climb but contained a photo of climbers with the figure caption "Tourism Increasing".[120] A consultant's report quoted in the plan indicates 87% of tourists came to see the Rock and 71% to climb. It seems NPWS were well aware why people were coming in increasing numbers – they were coming to climb.

The 2000-2010 management plan has an increasingly sanctimonious tone and says little about non-Anangu cultural heritage at The Rock. The plan is subtitled *Tjukurpa above all else*, the ideology of a small group now to be the focus of the Park rather than its natural attractions. It's almost as if the Custodians are jealous that their Rock gets all the attention. Early exploration is given scant treatment in the plan. There is no mention of the Horn Expedition for instance and no mention whatsoever of the ground-breaking anthropological work of Charles Mountford. Scientific information about the park's geology, flora and fauna are lacking detail and are subordinate to Anangu myth and superstition – this is also reflected in Park signage. Sadly, the long tenure and dedication of former Head Ranger Derek Roff is completely overlooked (as it was in the 1986 and 1991 plans). History is being deleted from the Park. In regard to the Climb the aim of the plan is to discourage people from climbing as it shows *disrespect for the spiritual and safety aspects of Tjukurpa*. Provision was made for a review of climb management three years into the plan *to gauge the success of the work done to discourage climbing and to identify alternative or additional measures that may be necessary including improved ways of telling visitors about why climbing is inappropriate, better ways of managing climbing or possible future closure of the climb.*

Access to the Climb was now also tightly regulated and controlled. A fence and gate were constructed at the base in the early 2000s along with a large sign featuring a "We don't Climb message". Closure protocols included the following measures, still current today:

For your safety the climb is always closed:

OVERNIGHT - from 5.00 pm.

SUMMER - from 8.00 am during the summer seasonal closure period: December, January and February.

The climb may also be closed with little or no notice because of:

HEAT - if the actual temperature reaches 36°C or above.

RAIN - when there is greater than 20 per cent chance of rain within three hours.

THUNDERSTORMS - when there is greater than 5 per cent chance of thunderstorms within three hours.

WIND - if the estimated wind speed at the summit reaches 25 knots or above.

WET - when more than 20 per cent of the rock surface is wet after rain.

CLOUD - when cloud descends below the summit.

RESCUE - during rock rescue operations.

CULTURE - if the traditional owners request closure for cultural reasons, for example during a period of mourning.

The protocols effectively mean the Climb is closed about 80% of the time. The winter months, particularly June, July and August, provide the best opportunity for meeting the gate with a "Climb is open" sign.

The 2010-2020 management has little mention of non-Anangu cultural heritage in the park, and there are no provisions for preserving, managing and maintaining that heritage as set out in the lease agreement. Section 17(2) of the lease agreement provides important protections for Anangu and Non-Anangu Cultural heritage. These include the Parks collective cultural history including the Climb, and its physical heritage including the chain, summit pedestal and memorial plaques. This section of the lease reads:

The lessee covenants that the flora, fauna, cultural heritage, and natural environment of the Park shall be preserved, managed and maintained according to the best comparable management practices established for National Parks anywhere in the world or where no comparable management practices exist, to the highest standards practicable.

A breach of this clause should effectively end the Government's involvement in the Park and in the future it can operate as a private park under direct Anangu control without taxpayer funding. At that point perhaps the locals will install a pillory at the base of the Rock as a deterrent to future climbers.

When the prospect of a ban was revealed in 2009, in a rare act of bipartisanship both then Prime Minister Rudd and Opposition leader Turnbull (ex-PM at time of writing) criticised moves to ban the Climb.[121] In a press release then Shadow Environment minister Greg Hunt proclaimed that Big Brother was coming to Uluru and that Minister Peter Garrett would be known as the "Minister Who closed the Climb". Despite the strong political opposition to the proposed ban, and the strong support for the Climb by the Australian public, the Australian Parliament still passed legislation to enact the plan without excluding the closure conditions. This is modern democracy at work: *Government of the Bureaucracy, for the Bureaucracy, by the Bureaucracy!*

Liberal Party of Australia

Wed, 8th July 2009

RUDD MUST NOT CLOSE THE ULURU CLIMB

The Hon Greg Hunt MP
Shadow Minister for Climate Change, Environment and Water

Kevin Rudd must veto any plans by Peter Garrett to shut down Australia's world-famous Uluru climb.

Plans to close the climb have been outlined in a draft management plan for Uluru released today by Peter Garrett's office.

Under the Garrett plan, visitors from around Australia and the world would be stopped from completing the majestic and exhilarating journey.

At a time when Australia's tourism industry is facing massive challenges from the global financial crisis, and the 350,000 annual visitor figure at Uluru is at risk, how can the Rudd Government even contemplate such a move?

The Prime Minister cannot allow Peter Garrett to go ahead with his plan to close the climb.

I have always suspected that closing the Rock to walkers was on Labor's agenda. Today we see the start of their plan to end one of the great tourism experiences in Australia.

Indeed, it was a Coalition Government that provided $20 million for a world-class viewing platform, but it was a Labor Government that is planning to take away people's choice.

Big Brother is coming to Uluru to slam the gate closed on an Australian tourism icon, the climb.

Peter Garrett would forever be the Minister who Closed the Climb.

Kevin Rudd would be the Prime Minister who shut down a sensitively-maintained, world-class tourist attraction in the midst of an international financial crisis.

Uluru is an indigenous treasure. It is also a national and an international treasure.

I support allowing people to make up their own minds about whether to make the climb.

It is a matter of enabling 'informed consent' – providing people with relevant information about cultural sensitivities, weather conditions and potential risks and allowing them to make the decision as to how best to honour and experience Uluru.

Figure 32: Greg Hunt Press Release: Big Brother is coming!

The conditions for banning the Climb were outlined in section 6.3.3(c) and included the following:

1. the Board, in consultation with the tourism industry, is satisfied that adequate new visitor experiences have been successfully established, or

2. the proportion of visitors climbing falls below 20 per cent (see the 20% myth), or

3. the cultural and natural experiences on offer are the critical factors when visitors make their decision to visit the park.

None of these conditions has been met, yet on 1 November 2017 the Board announced the Climb would be banned from 26 October 2019. A new National No-Go zone is to be born!

So why are the Board banning an activity that has brought so much joy to so many? Sammy Wilson, chairman of the park board, gave this reason:[122]

The Climb is a men's sacred area. The men have closed it. It has cultural significance that includes certain restrictions and so this is as much as we can say. If you ask, you know they can't tell you, except to say it has been closed for cultural reasons.

What does this mean? You know it can be hard to understand – what is cultural law? Which one are you talking about? It exists; both historically and today. Tjukurpa includes everything: the trees; grasses; landforms; hills; rocks and all.

Parks Australia indicate that *Tjukurpa* never changes.[123] If so, then we are left to wonder how the views and actions of past elders like Paddy Uluru, Toby Naninga, Tiger Tjalkalyirri and others have come to be ignored and apparently disrespected by the current board. If Tjukurpa never changes then their views about access should be respected by the current custodians. Paddy Uluru indicated the Climb had no cultural significance, Toby Naninga said that aside from the Ngaltawata Pole and Warayuki tourists could go anywhere else, Tiger was one of the first climbing guides, his actions an example of Tjukurpa in action. Australian's were assured by the Traditional Custodians in 1983 that tourism and access would be business as

Figure 33: Current sign at the base of the climb (Source Wikipedia[124])

usual. Who would know better what *Tjukurpa* said about visitors climbing the Rock? The men who lived and practised traditional ways, or the group that came after them with a different agenda?

Visitor Numbers 1958-2017

The financial year 2000-01 recorded the highest visitation with close to 400,000 adults passing through the gates, with over 70% of those climbing the Rock – the numbers likely bolstered by the flow-on effect of the Sydney Olympic Games. Children did not start paying until 2016 so Parks' figures, prior to this, under-report the total number of visitors in the order of 10% hence we have adjusted earlier figures by this amount. Since 2000 visitor numbers have decreased dramatically, coinciding with stronger emphasis in marketing material of the Aboriginal ideology "Tjukurpa" including the plainly false "We never climb", "Do Not Climb" message, increased regulation of photography, less access, more signs and tighter closure protocols at the Climb. These effectively close it 80% of the time. In 2017 there were 303,015 paying visitors, including children over the age of five, a drop of over 30% on 2000-

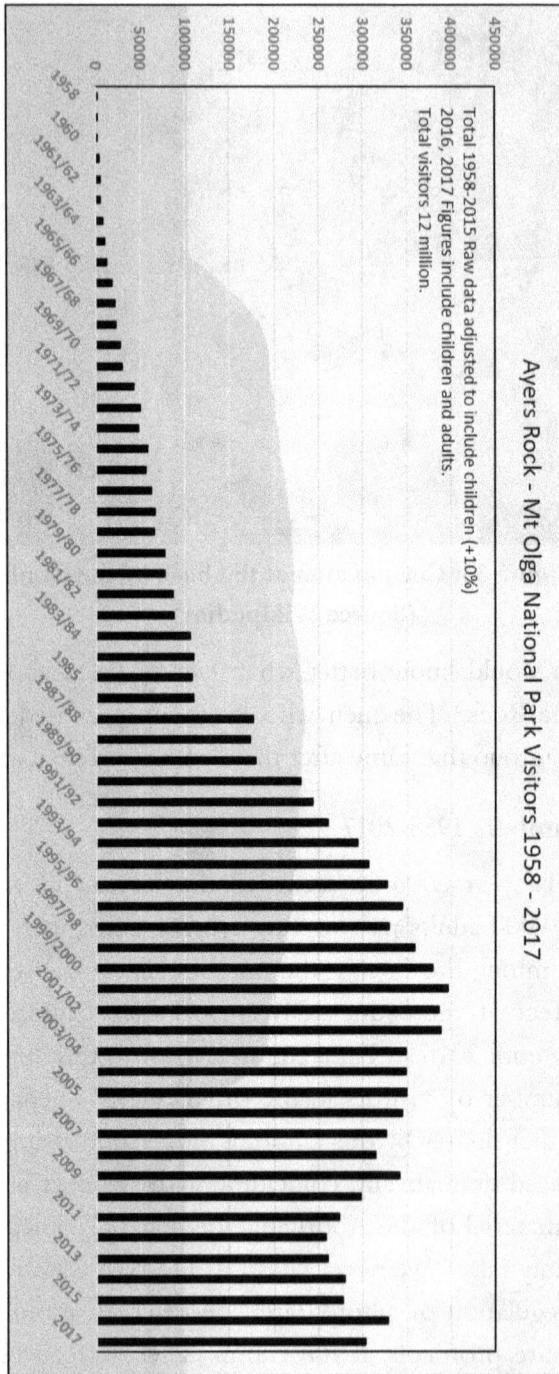

Ayers Rock - Mt Olga National Park Visitors 1958 - 2017

Total 1958-2015 Raw data adjusted to include children (+10%)
2016, 2017 Figures include children and adults.
Total visitors 12 million.

Figure 34: Ayers Rock-Mt Olga National Park. Annual Visitors 1958-2017 based on Parks Australia and other sources

01 figures. Given the ban on climbing goes ahead, the number of visitors will likely fall further. Based on climb data (see below) we expect visitor numbers will fall by at least another 30% once the ban comes into force to about 200,000. Any uplift in numbers up to October 2019 will be due to an influx of people wishing to experience the grandeur of the Climb before it gets the chop.

Number of Climbers 1958-2017

The proportion of visitors climbing has changed significantly, reflecting changes in management attitudes and regulations and not visitor intentions. Prior to the installation of the chain the proportion was about 20%, which increased to close to 80% after the chain was installed. Numbers only began to be affected by the dramatic change in policy, and introduction of conservative closure protocols and negative anti-climb propaganda by Parks Australia in the early 2000s. A statistical analysis of climb counter data (see Chapter 7) indicates the proportion of visitors climbing still reaches beyond the 70% mark when the Climb is fully open from sunrise to sunset, with the average proportion on those days sitting at 44% over the period 2011-2015. The total number of climbers is in the order of seven million or about 60% of all visitors. This does not

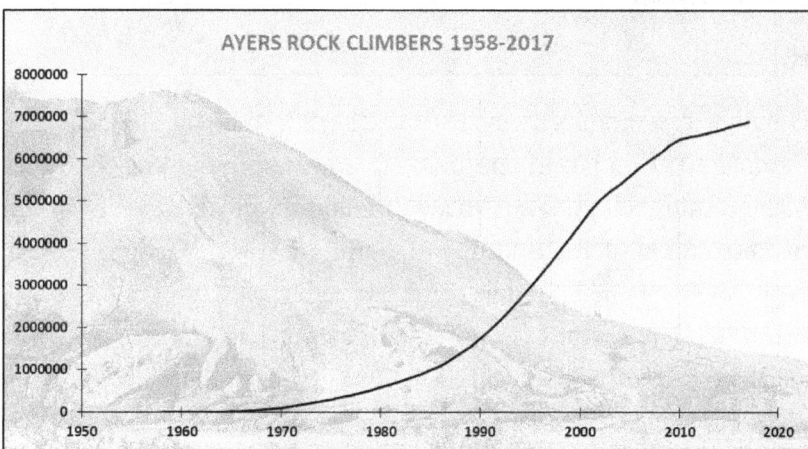

AYERS ROCK CLIMBERS 1958-2017

Figure 35: Ayers Rock-Mt Olga National Park cumulative climbers 1958-2017. Total for the period is nearly seven million

Figure 36: End of an era. A young Craig Woon at the top of Ayers Rock June 1986. Logbook was still in place, perhaps the last one (Photo courtesy Craig Woon)

take into account the fact that some visitors climb more than once. This is a remarkable figure reflective of a remarkable experience.

Future

The Board's announcement that the Climb is to be banned from 26 October 2019, and its intention to effectively strip the Rock of non-Anangu cultural heritage, is an act of cultural vandalism, on par with the destruction of the Buddhas of Bamiyan. It is difficult to see the ban being sustained for long and ultimately the Climb is likely to return in some form. Like other "attractions" at what has become a sideshow alley version of Disneyland, it will probably come with a high price tag.

The independence and freedom experienced by earlier visitors has been relegated to history.

2

AYERS ROCK GEOLOGY

When you stand on the top of Ayers Rock you are standing on a pile of sand and grit that formed about 550 million years ago and was tilted 80 degrees and fully lithified during dynamic mountain building events 450 to 300 million years ago. It has taken over 100 million years for the forces of erosion to slowly carve the natural wonder that lies beneath your feet. Understandably as you stand up there and look at the remarkable scenery, you feel a sense of awe.

Early visitors to the Rock took the lithology for granite, including Gosse and Tietkins. Baldwin Spencer was the first to correctly identify it as having a sedimentary origin. His 1895 diary notes record the following observations: *Ayers Rock consists of rock make up of rock compound of water iron fragments of Quartz, Felspar. Mica with cementry material containing a mineral matter –probably magnetite.*

The brief geological description below is based on the 2002 Ayers Rock 1:250000 sheet explanatory notes,[125] and the excellent *Uluru and Kata Tjuta: a geological guide* published by Geoscience Australia[126] to which readers are referred for a more in-depth geological lesson.

The formal stratigraphic name for the Rock is the Mutitjulu Arkose, considered to have a late Neoproterozoic to early Cambrian age. It was formed from sand and gravel washed into a broad alluvial basin by energetic rivers flowing away from mountains formed by the Petermann Orogeny (between 550-530 Mya) to the south and southwest. The formation has been considered as a lateral finer grained variant of the Mount Currie Conglomerate (that forms Kata Tjuta), but recent interpretations based on the results of geophysical surveys suggest it is probably more likely to represent a separate alluvial fan. The Mutitjulu Arkose has been equated to the Arumbera Sandstone in the northern Amadeus Basin. The exposure at Ayers Rock is steeply (80-85°) southwest dipping and youngs the same way, the older beds are on the north side.

The sedimentary origin of the Rock is clear from the presence of bedding, which has given rise to the distinctive large furrows or grooves on the sides and summit due to differential erosion of layers with slightly different properties, and well preserved sedimentary features including multidirectional crossbedding, scour and fill troughs and ripples. These features indicate deposition in a high-energy fluvial (terrestrial river) environment of broad shallow channels or sheet floods. Similar environments today are found in alluvial fans forming at the foot of major mountain ranges. Four rock facies, based on lithology, grain size and sedimentary structures have been identified:

- well sorted, planar bedded or ripple laminated, very fine to medium sandstone;
- trough crossbedded sandstone to granule conglomerate;
- planar bedded and crossbedded granule to pebble conglomerate; and
- massive conglomerate.

Cross bedding, including trough cross bedding is best observed in the cave Kulpi Watiku,[127] also called Malaku Wilytja,[128] or the sound shell, accessible via the base walk on the west side of the Rock but is well preserved where fresh faces are exposed. Pebble and granule conglomerates are found throughout the exposed rock face. The rock is best characterised as an arkosic arenite. It typically comprises up to 50% feldspar grains, 25-35% quartz grains and up to 25% rock fragments. The lithic rock fragments include basalt and rhyolite. Clasts and grains generally range from 2-4 mm and are set within a matrix of polycrystalline quartz. The strong crystalline cement contributes to the Rock's durability. The high proportion of course, subangular feldspar grains (K-feldpar dominates over plagioclase) gives the Rock a "granitic" appearance and along with the massive unbroken nature of the outcrop is what probably confused early explorers.

The Olgas or Kata Tjuta 30km to the west comprise spectacular shallow (15°) southwest dipping layers of pebble, cobble, and boulder conglomerates. They form the Mount Currie Conglomerate, which unconformably overlie the Winnall beds at Mount Currie 35km to the northwest. Young et al (2002) identified three informal members based on the composition of the clasts:

- A basal member with distinctive sandstone clasts, thought to have been derived from the underlying Winnall beds.

- A middle member comprising a polymictic (mixed clasts) conglomerate with an igneous provenance. The member is dominated by clasts of porphyritic rhyolite (70%), basalt (20%) and about 10% sandstone clasts. The Walpa Gorge is a good example.

- An upper member that has more granite than rhyolite clasts.

The change in the composition of pebbles and boulders is due to progressive uplift in the source area exposing deeper and different rocks to weathering.

Following deposition both formations (Mutitjulu Arkose and Mt Currie Conglomerate) were covered by younger formations in the Amadeus Basin and slowly turned to solid rock. The wonders of that overlying geology are beautifully laid out in the West MacDonnell Ranges to the north. Sedimentation in the basin ended during the Alice Springs Orogeny about 340Mya. This was another significant mountain building event in central Australia. Deformation at the Olgas was relatively mild, resulting in moderate tilting of the strata and jointing, while the Mutitjulu Arkose was folded into a tighter structure. The exposed area at Ayers Rock preserves part of the southern limb of an anticline.

A long period of erosion followed the Alice Springs Orogeny lasting to present day and for virtually all the time the area that is now the Park has been above sea level, aside from a brief period during the Cretaceous. Ayers Rock, The Olgas and Mt Conner are remnants of an extensive flat lying landscape that formed in the Late Mezozoic, 100 million years ago.[129] The enveloping rock has been progressively weathered away leaving behind these three great inselbergs. An analysis of the weathering history of Uluru by Charles Twidale suggests the summit surface is probably a 70 million-year-old remnant of that extensive Late Mesozoic plain. Cretaceous sediments found in water bores around Ayers Rock indicate there was a prominent hill here at that time. The formation of Ayers Rock is a story of how a slight topographic rise was slowly converted into the prominent steep sided inselberg we see today through progressive differential weathering, subsequent erosion of its flanks and episodic lowering of the surrounding plains throughout the Cenozoic.

The major erosion agents have been water and time. Due to the lack of water and homogeneity of the outcrop, erosion rates in this desert environment are very slow. The deep valleys at the Olgas were slowly carved out parallel to vertical joints that developed during the Alice Springs Orogeny. These areas of weakness are more prone to weathering and erosion than the adjacent Rock which is remarkably free from tectonic jointing. As you climb up mostly along the same group of beds till you reach the top of the chain you will not see a single joint or other structural defect crossing your path. However there are "unloading" joints present that parallel the rock surface. These develop in reaction to progressive unloading of the ground surface. A number of these have given rise to important cultural features such as the Ngaltawata Pole (Kangaroo tail). This is the remains of a once continuous sheet of rock that split from the main mass millions of years ago. Most of it has collapsed leaving a stout column of rock behind. Another similar feature is observable from the climbing spur mid-way up the chain if you look to the south. Exfoliation of very thin layers of rock (up to 10s of centimetre) occurs on the sides and summit due to the effect of water penetrating shallowly into the rock surface causing it to lift off in thin sheets, combined with temperature extremes that contract and expand the surface. Much of the rock surface sounds hollow due to this process. Lightning strikes on the Rock cause explosive exfoliation to take place, as water under the surface is rapidly heated and turns to steam. Smaller caves and erosional features are due to the effect of salt crystals dissolved into percolating rainwater that migrates downwards and outwards to the surface where it causes volume expansion and fretting when the salts crystallise out. Larger caves are initiated by longer term processes tied to previous ground water levels and subsequent exposure. As caves develop, seepage acts to progressively widen and deepen them creating overhangs. When the weight of the overhanging rock is greater than the rock's strength these collapse and the process begins again. It will take many tens of millions of years for these processes to wear away these rocks.

The extensive sand plains surrounding the tors were formed during the last ice age between about 40,000-16,000 yBp. Dunes were likely still active when the first humans reached the Western MacDonnell Ranges (Cleland Hills) 160 km north of Ayers Rock about 30,000 years ago. Central Australia was likely a much drier and less hospitable place then, with much less standing water available. It would have been even more difficult for humans to survive in that environment.

The dunes are mainly irregular and discontinuous (reticulate), however some longitudinal dunes of several kilometres are known from the Ayers Rock region. The dunes exhibit an overall south-easterly trend and may attain a height of 15m. Small shrubs have stabilised most dunes, whereas the sand plains are populated by a variety of shrubs, small trees and spinifex.

The red colour of the sand and weathered surfaces in the Park are due to oxidation of iron in the rock to form iron oxide, or rust. Colours are accentuated at sunrise and sunset by the sun's low angle rays that are at the redder end of the spectrum.

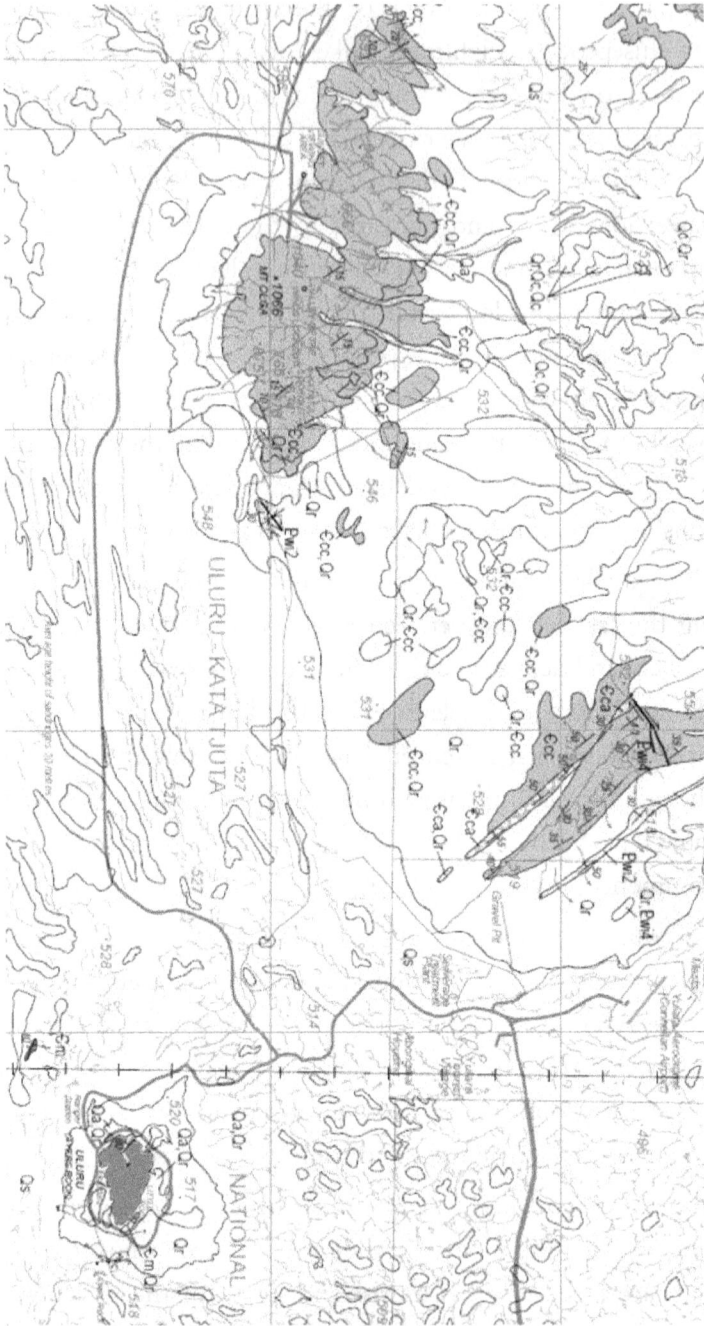

**Figure 37: Geology map from Ayers Rock 1:250000 sheet, €cc
Mount Currie Conglomerate, €m Mutitjulu Arkose
(Reprinted with permission Northern Territory Geological Survey)**

3

REASONS TO CLIMB AYERS ROCK

- **Exhilarating physical experience**

It's difficult to put into words the feeling you get when you reach the summit of Ayers Rock, and climb back down. My youngest daughter looked for words to describe her feelings about her climbing experience. Coming down from the summit the second time, she asked me, "What's that word that means feeling alive?" And I replied "Do you mean 'exhilaration'?" "What's that?" she asked. "Sort of a feeling of bliss, excitement, happiness, joy, and accomplishment all mixed together." "That's what I feel", she said, "Exhilaration! I feel alive".

- **Views from the summit are extraordinary**

Most visitors to the summit agree that the views are extraordinary and well worth the walk up. Some say all you can see is the desert. Some people also say Botticelli's *Birth of Venus* is just a naked lady standing in a shell, or that Van Gogh's *Starry Night* is a mess of blue swirls that doesn't depict the night sky that well. To better appreciate the subtle beauty of the desert and mountain ranges laid out before you at the summit, an understanding of the processes that led to its formation help to appreciate its grandeur and the importance it has to human settlement and culture. One of the wonderful things about the views are the changing colours with the day and season. Climbing during the day, the Olgas are lit up bright orange to the west, but later in the day take on more purple tones as the sun sinks behind them. The patchwork mosaic of vegetation among the sand ridges, reflecting different soils and geology and the faded colour of distant ranges visible on the horizon, combine to make the views among the best in central Australia.

- **Low risk activity for fit and healthy people**

Over seven million people have climbed the Rock. Even using the more conservative figures of Parks Australia of 37 deaths the rate of fatalities is well within the range for similar activities elsewhere in the world. For fit and healthy people who keep to the marked trail the risks are quite low, less than two micromorts per climb.

- **Best way to fully appreciate the surrounding landscape, geology and geomorphology**

The Climb provides a means of linking all the physical features into a coherent story. Looking across the surrounding plains from the summit climbers are able to fully appreciate the vast time scales required to have left it and The Olgas prominent above the plains, while adjacent rocks have been worn away, in a way that you just don't get from walking around it, or viewing it from a distance. The physicality of the Climb and close connection to the Rock makes it all the more real.

- **Source of inspiration, awe and wonder**

Both Anangu and non-Anangu see the Rock as a source of inspiration and wonder. Its majesty and beauty prompting great works of art in every field, from writing to paintings, music, dance and science.

- **Fundamental part of human culture**

Climbing things is an integral part of human culture. We climb to seek a vantage point from which to look out for water, food, predators and rivals. We climb for ceremonial reasons, perhaps to get closer to heaven, to gain power, to lift ourselves above the mundane, as an act of freedom of expression, exploration, curiosity and physical adventure. We climb for science, to understand the natural world and how it works, and to find clearer skies to view the wonders of the night. We enjoy exploring new areas and seeing with our own eyes the world's natural wonders. Where there is no chance to climb we have built things to climb, like the pyramids, and in the modern era, skyscrapers and towers. Climbing is fundamental to our being.

- **Integral to understanding Aboriginal culture**

There are a number of Anangu creation myths that include features on the summit.[130] These are away from the main climbing route. If you seek to appreciate Anangu culture then you need to add the Climb to a trip around the base. It's a pity there are no guided tours to culturally significant areas of the summit plateau.

- **Respects the views of past owners who also climbed**

Climbing respects the views and actions of past Traditional Custodians, like Paddy Uluru and Tiger Tjalkalyirri who also climbed. Neither man indicated the act of climbing disrespected Anangu culture.

- **Builds relationships and leaves lasting memories**

Many climbers describe the camaraderie of the Climb that helps connect them together and builds lasting memories – the good natured greetings made to climbers going up and down; the shared emotions felt exertion and fear; the euphoria felt on reaching the top. These feelings are much more intense than those you'll experience undertaking any other activity in the Park. The base walk is worth doing but does not have the same intensity.

- **Because you should not feel guilty about enjoying the natural world.**

Everyone should be free to make up their own minds and enjoy the natural world without being made to feel guilty or be subject to racial, religious or political inference.

- **Source of scientific inspiration.**

My path to the geological sciences was formed by many stepping stones, among them visits to various geological wonders at a young age. Growing up in Victoria this included Hanging Rock, Organ

Pipes, various old gold mines, the granites at Mt Buffalo, caves at Buchan and the Twelve Apostles. From where will future geologists get their inspiration when so many natural wonders that inspired us are being closed to the public?

- **There to be shared**

The Rock is there to be shared. According to the local myths the route of the Climb marks the path of the Mala men on their arrival at Uluru at creation time. Other religions celebrate and share their stories and places of worship openly with the rest of the world. Visitors to the Vatican may walk through the cathedrals of Rome appreciating their artistic and architectural beauty without requiring an indoctrination in Catholicism. Virtually every church tower of substance is open for visitors to climb and enjoy without having to pay homage to a God; you can just go and enjoy the views. Visitors to natural wonders in the state of Utah are not required to attend a Mormon church service. You can climb the Atlas Mountains in Morocco and you won't need to tip your hat to Mohammad. Climbers of Mt Everest are not required to pass a religious exam. Not all visitors are interested in learning tribal customs or traditional laws. Most go for the view, for the thrill, to experience natural wonders in all their glory. At Ayers Rock it seems visitors are only truly welcome if they pay homage and adhere to the belief systems of the Traditional Custodians. Some equate climbing the Rock to walking on the roof of a cathedral, but when you climb the Rock you're not walking on top; the Climb is actually the path inside. Rather than lock tourists out Traditional Custodians should embrace the Climb as a means of celebrating positive aspects of their culture and share it with others, as they do with the interpretive walks around the base of the Rock. The story of the Anangu forms part of our collective human journey. This is where we came from; but through curiosity and a search for rational explanations about the world around us, look at what we have become, look at what we have achieved. The Climb is something to be shared with the rest of the world.

4

Ayers Rock Summit Routes

Park Authorities have never been enthusiastic about visitors using routes to the summit other than the main climbing spur on the western side, the only legal way up. However, for experienced walkers and climbers there are a range of options to reach the summit. In the 1960s Pioneer bus driver, Graham Phillips had reportedly climbed the Rock in seven different places. He was very fit and ran four miles before breakfast from the Pioneer Lodge Motel to the Rock and back again.[131]

Main climbing spur-chain route

The steep spur at the western most part of the Rock (see the cover) is certainly the easiest route up, and Arthur Groom described it in 1947, well before the chain was installed, as *nothing else but a strenuous and spectacular uphill walk*.[132] He also noted at least two places on the southern side where routes likely involved *a rough and steep scramble up*. Mind you Groom was a man who thought nothing of walking a hundred kilometres a day, he was incredibly agile and fit.

In his book covering his explorations in central Australia, *I Saw a Strange Land*, Groom, 42 at the time, gave a wonderful description of walking up the west side with his Aboriginal guides, Tiger Tjalka-lyirri, Tamalji and Njunowa (see history). The three men provided a representation of different climbing styles, with Groom and Tamalji racing ahead of a persistent Tiger, with Njunowa stopping near the base, his actions exemplifying the notion that it is a wise man that knows his own limitations. The stopping point is known as Chicken Rock. Prince Charles and Lady Diana climbed to this point when they visited in 1983.

The relative ease of the main route, especially after the chain was installed in 1964, accounts for the success of the Climb here and the reason it is so popular with visitors. It is daunting to people with little bushwalking or climbing experience, but achievable to most people of average fitness who are able to deal with the feeling of exposure. The average grade along the steepest section with the chain is 46% or about 25°, though short sections are a little steeper. By comparison stairs range between 20° and 45°, but typically between 30° and 38°.[134] Experienced walkers will climb this way unassisted, typically spurning use of the chain. The point along the upper section of the chain, where the spur narrows with steep drops on both sides, tests the nerves of many. The long descent is made by many by sliding on their backsides.

Over seven million people have used this route, many following it more than once. Immediately around the chain, no more than one metre wide, and along the painted path across the summit plateau the rock has worn smooth, making the surface slippery in places, especially when wet. Away from the marked route, however, the rock surface is still rough and provides a better walking surface.

Figure 38: View looking up the main climbing spur along the steepest section (Photo Dana Hendrickx)

Average climbing time to the summit and back again via the chain route is about one-and-a-half hours.[135] Overall time depends greatly on for how long you linger to take photos of the glorious views, and how long you choose to soak them in and rest at the summit. Remarkably, the fastest times up are below 13 minutes, and down, below 11 minutes.[136]

Figure 39: View looking down the steep section of the main route. The worn path either side of the chain smoother and more reflective than the adjacent rougher surface. The action of over 14 million feet has had very little impact and is confined to a narrow zone next to the chain (Photo Zoe Hendrickx)

Alternate routes

There is currently only one legal route to the summit, and any tourist found wandering off the marked trails in the Park faces significant fines if caught. Other ways up the Rock are possible and have been undertaken in the more enlightened past.

The routes along the southern side of the Rock mentioned by Groom are likely to be in the steep slopes above Mutitjulu water-hole, Metjan Rockhole and Tjukiki Gorge (see following photo and map). These occur in valleys and, despite being slightly steeper on average than the main climbing route, are arguably safer options,

Figure 40: For comparison with Figure 39 this similar view without the chain taken slightly higher on the spur in 1952 by Keven Harris (State Library of South Australia B 70782/75)

as climbers are not as exposed to potential falls from near vertical cliffs. If they lose grip and fail to stop, however, there would be a long and uncomfortable slide down the equivalent of a cheese grater to contend with. The absence of any support means these alternate routes should be attempted by experienced walkers only. Average grades for the steepest sections range from 45% through to 56% for the shortest route that starts at Metjan Rockhole next to the Ikari, or smiley cave (see Figure 42). There are likely to be other possible routes on the west side of the Rock between the climbing spur and the Ngaltawata Pole; but these are difficult to access from the very toe and may require some technical climbing to get a start. A possible exception is the steep slope above the Kandju Cave on the northern side of the Kandju Soak.

The Mutitjulu Route was successfully negotiated by Cyril E. Goode in 1952, Peter Magnus and Milton Osborne to name a few. In an "Account of a Dash to Ayers Rock",[137] Goode recounts: *Left*

Figure 41: Possible climbing routes southern side of the Rock. MJ Mutitjulu, MR Metjan Rockhole, TG Tjukiki Gorge (Photo by J Fitzpatrick, 1957, National Archives of Australia BC 11927525)

to my own resources I began to explore the surface of the rock to the right and noticed the set of steep ridges pitching down into the great fissure that gives rise to Maggie Springs. Finding the surface broken enough to climb with bare feet, enthusiasm gripped me and I found myself going up and up, gaining seventy or eighty feet on every ridge. There are seven ridges, and in this way the summit is gained without any strong sense of the horror of falling. One appears to be climbing in precipitous valleys with always a number of water-falls or basins below except when crossing the ridges, and then it is best not to look down. I had forgotten to take chalk to mark the descent, and about half way back seemed to be going wrong, so the others directed me from the ground. My feet were bleeding long before the bottom was reached.

Good footwear is a must!

Less is known about the other routes along the south side, which were likely used by Allan Breaden in his traverse of the Rock, Graham Phillips and other climbers over the years. Bus driver and guide Arnold Derks is reported as climbing 14 times, once from the east side, possibly via Tjukiki Gorge.[138]

A possible sanctioned traverse across the Rock, in the footsteps of Breaden, perhaps up from Tjukiki Gorge, via the Uluru Rockhole then across the summit and down via the chain route would provide a spectacular addition to walking options at the Park. With visitors making better use of the base trail to walk between the start and end points. The potential for additional via-ferrata style routes on steeper sections of the Rock and also at Kata Tjuta would greatly expand the range of visitor options. Perhaps future owners will see the merit?

Technical climbs

The only technical climb of the Rock we are aware of was done by experienced rock climbers Keith Lockwood and Andrew Thomson. In May 1973 they successfully climbed up the Ngaltawata Pole (Kangaroo Tail) on the north western corner.[139] They had attempted this the previous year but after being noticed were chased away at gun point by the Ranger (probably Derek Roff). The feat was repeated later in the 1970s and is something only a handful of people have accomplished.

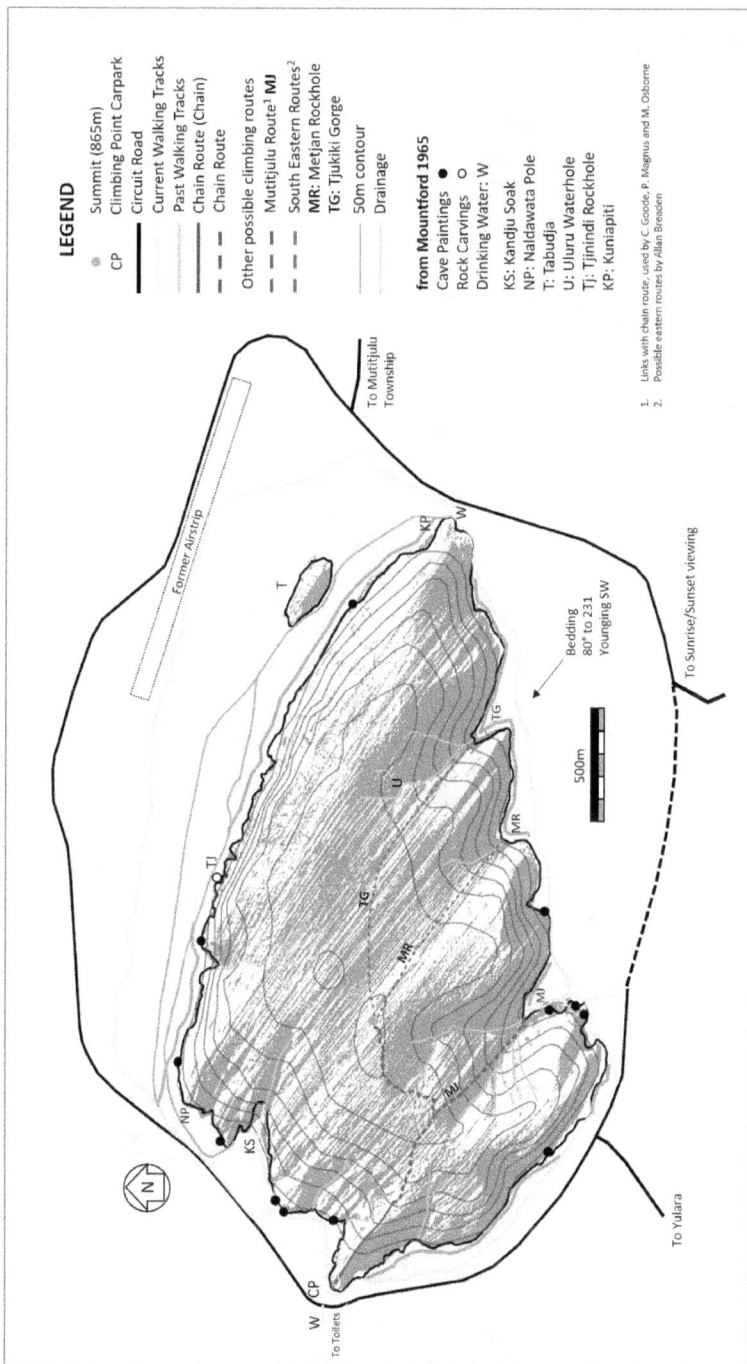

LEGEND

- Summit (865m)
- **CP** Climbing Point Carpark
- Circuit Road
- Current Walking Tracks
- Past Walking Tracks
- Chain Route (Chain)
- Chain Route

Other possible climbing routes
- Mutitjulu Route[1] **MJ**
- South Eastern Routes[2]
- **MR:** Metjan Rockhole
- **TG:** Tjukiki Gorge
- 50m contour
- Drainage

from Mountford 1965
- Cave Paintings ●
- Rock Carvings ○
- Drinking Water: W

KS: Kandju Soak
NP: Naldawata Pole
T: Tabudja
U: Uluru Waterhole
Tj: Tjinindi Rockhole
KP: Kuniapiti

1. Links with chain route, used by C. Goode, P. Magnus and M. Osborne
2. Possible eastern routes by Allan Breaden

Figure 42: Map showing topography and major features of Ayers
Rock, along with walking trails and climbing routes

91

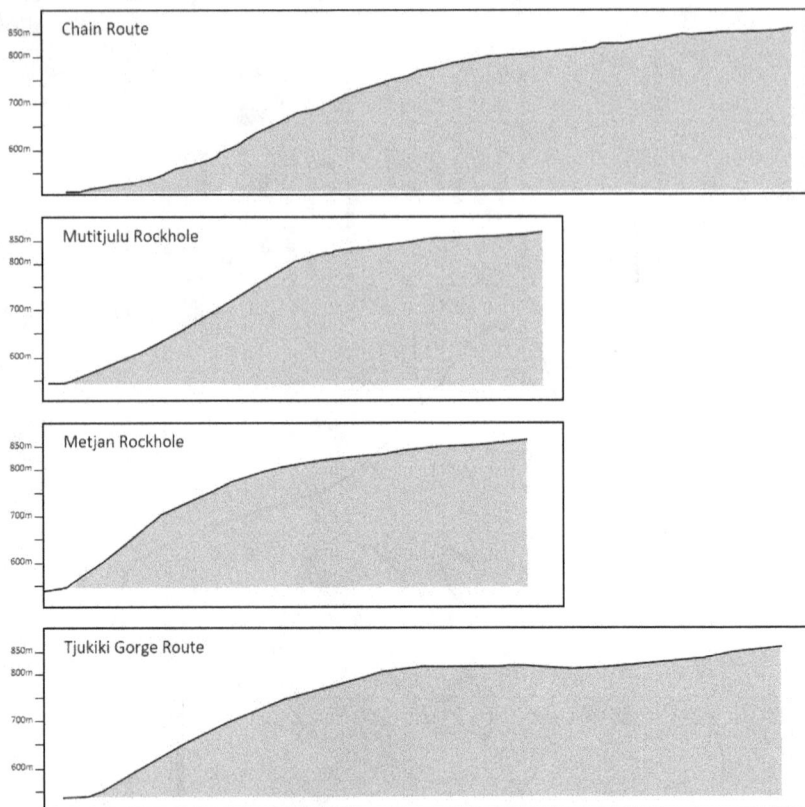

Figure 43: Indicative profiles of selected climbing routes up Ayers Rock

5

SOME PRACTICAL ADVICE
ABOUT THE CLIMB

Since Parks Australia and the Park Board began discouraging climbing in 1991 the level of practical advice, hints and tips about climbing has diminished to next to nothing. This section aims to provide some practical advice to climbers.

When to climb

Winter is the best time to climb for a host of reasons. It's cooler, drier, there are fewer flies about and the Climb is more likely to be open (check your odds below; August is the best month). On the downside it will be busier and prices for accommodation at the resort are generally higher. During winter, wind is most likely to be the factor that will see the gate closed at the base, rather than rain or temperature. Afternoons tend to be calmer, so if the Climb is closed in the morning, there is a good chance it will be open after lunch or mid-afternoon. If you find a ranger at the base and the Climb is closed don't be afraid to ask why, and whether it's likely to be open later in the day, or open the next day. If you want to climb make sure you drop past the climbing car park first thing in the morning and no later than 3pm in the afternoon.

Under the current closure protocols you'll only get to climb in summer if you are very lucky to find it open briefly first thing in the morning, or you are prepared to break the rules. It's also very very hot, even in the early morning and into the late evening. The flies are terrible. We suggest a beach might be a better option than going to Ayers Rock in summer, especially if your intention is to climb it.

Month	Jan	Feb	Mar	Apr	May	Jun	Jul	Aug	Sep	Oct	Nov	Dec
Ave Fully Open days	0	0	1	3	6	5	5	9	4	0	0	0
Ave Fully closed Days	22	23	19	10	9	9	7	6	10	18	22	24

Average number of fully open and fully closed days 2011-2015. August the highest probability of the Climb being open. Statistics based on data provided by Parks Australia through a Freedom of Information Request.

Will it rain?

Monthly rainfall data from the Bureau of Meteorology unsurprisingly shows the dry season to have the lowest rainfall. Another reason winter is climbing season!

Month	Jan	Feb	Mar	Apr	May	Jun	Jul	Aug	Sep	Oct	Nov	Dec
Ayers Rock (15537) 1964-1983	48	46	50	25	22	21	9	13	19	24	35	19
Yulara (15635) 1983-2018	29	38	33	16	14	18	18	5	9	21	33	46

Mean monthly rainfall for Ayers Rock and Yulara 1964-2018 in mm rounded to nearest mm. Station moved from the Rock to the Airport in 1983, numbers refer to BOM stations. Data from Bureau of Meteorology website.

What to wear

For both winter and summer, shorts and a lightweight shirt will generally suffice. The level of sun protection is up to you. In winter, a light jacket may come in handy stowed in a back pack if you spend some time at the summit, as the winds can be quite fresh. If you wear a hat make sure it has a strap, or can be tightened to your head. If it blows away you may not be able to safely retrieve it.

The last time we climbed in July 2018 I wore lightweight pants (Kuhl Renegade Stealth), and long-sleeved shirt (Craghoppers long-sleeved shirt). Daughters wore t-shirts and shorts or running pants.

Footwear

Good fitting rubber soled shoes with good grip are essential, and we also suggest boots with good ankle support to reduce the risks of spraining ankles, particularly when negotiating the ups and downs of the summit. If your shoes are too big you'll find your feet slipping in them on the descent and your toes will be quite sore and probably blistered by the time you get down. Don't climb in bare feet, thongs or shoes with no grip such as leather soled shoes, or high heels. On our last climb our daughters wore running shoes and I wore Scarpa Terra GTX walking boots, the latter were well worn but had enough tread to provide decent grip up and down.

Backpack and small items

A small backpack will provide a useful stowage space for drinks and snacks, camera(s), phone and a place to store any rubbish. Be careful taking it off, especially on the steep section. At least one climber has fallen to his death attempting to retrieve a backpack that slid away from him. If it, or anything else you are carrying falls off the rock, and you can't safely retrieve it, then just leave it! Check with a Ranger once you are back down to earth to see if it can be retrieved. Depending on where it went down it's unlikely it will be found. Look out for pieces of broken cameras and other scattered materials at the base of the Climb near the five memorial plaques.

Food and drink

At least one death on the Rock has been ascribed to a large breakfast, eaten just prior to starting the Climb. This results in blood being directed to your stomach instead of places it's needed when exerting yourself. Just like going for a swim, don't gorge yourself before you climb. As you are unlikely to be up there that long, you really don't need to bring any snacks, but if you are the snacking type consider something like a muesli bar. Just don't leave the wrappers up there.

In winter I found that a litre of water per person provided sufficient refreshment for the Climb, with some left over. Drinking water is available from taps opposite the carpark. You will likely find it harsh on your palate as it has a strong taste of plastic. If you are particularly fussy make sure you have your own supply. The IGA supermarket at Yulara stocks ample supplies of bottled water.

How long will it take?

In the 1980s the average time up and down was reported as 1.5 hours.[140] The record time up is under 13 minutes and down, under 11 minutes. Two hours should provide ample time for the average climber without needing to over exert yourself on the way up, and provide time to look around the summit and at the remarkable desert views.

Toilets

There are no toilet facilities on the Rock itself but there are toilets located near the base of the Climb. Make sure you use them before you start climbing. Remember for most people it's going to take about 1.5 to 2 hours to go up and down, about the length of a feature film.

Don't leave things behind; please take them back with you! An empty drink container and/or a couple of plastic bags may be a handy addition to your backpack.

Climbing with children

Derek Roff wisely states:[141] *When Climbing with young children, insist they remain with the adults in the party. Children usually have less fear and more energy than their parents and need watching.*

How to climb Ayers Rock

Great deeds are accomplished by simply putting one foot in front of the other.

Fit and healthy people who have previously ventured up mountains or regularly bushwalk will have no problems climbing Ayers Rock along the main climbing route and won't need any advice. Others less experienced may find it more of a challenge, but one they can achieve if they are willing to take their time and persist. Some will struggle and perhaps turn around long before the end of the chain, and some will look at it and wonder why you'd bother? It's up to you; just remember that it is a wise person that knows their own limitations.

Once through the gate the route is quite obvious. The lowest part of the Climb is on a massive slab of sandstone that is separated from the main massif along a joint surface subparallel to the intact rock surface. This is quite broken up on the left hand side of the spur. The end of the broken rock marks "Chicken Rock". The start of the chain is less than 20m from here. It used to be a little higher, but it was extended down to this point sometime in the late 1970s or early 1980s.

Different people will tackle the short hike up to the start of the chain in different ways. Experienced walkers will simply walk up to the base of the chain and continue upwards without any reliance on it. Others go up on their hands and feet, bent over like crabs, or inch up backwards on backsides grateful to reach the chain to have something to hold on to for the rest of the steep climb up. There are 370m and 136 posts to the point where the lower section of the chain ends. In general most people keep to the left side up and down, but if the other side looks better don't be afraid to use it, just be careful stepping over the chain and passing others or when they pass you. The grade is variable but typically as steep as, or just a little steeper than a set of stairs. The rock surface right next to the chain is well worn and care is needed not to slip. On the occasions I have climbed, I have taken advantage of the rougher surface adjacent to the main path and used careful foot placement to walk up without using the chain, but this takes some skill and leaves you unprotected in case you slip or fall. About 250m along the spur it gets quite narrow leaving most climbers feeling more exposed. If you suffer from a fear of heights this is the point where you will most likely feel it the most, and if it's windy you might be inclined to grip the chain a little tighter as you make your way to the end.

Figure 44: Your journey starts here. View from the base of the Climb looking up (Photo Dana Hendrickx)

Approaching the end of the chain the grade reduces a bit and you'll find yourself in a small protected shelf with wonderful views to the Olgas to the west. You are standing about 160m above the plain, half way up the Rock with another 155 vertical metres to go to the summit spread over another 1000m. This next bit has an average grade of about 15%. A short section of chain here supported by two posts marks your entry to the summit plateau. The path from here is marked by a white painted line. Like the spur, the rock along the line is well worn in places and sometimes quite smooth. Better grip may be found by walking next to it. You can see where the rougher surface is, so take care not to trip on the irregular surface.

The walk over the shoulder to the summit plateau generally follows the crest of a few of the ridges for about 400m from the chain and then turns northeast at a spot where you will have good views looking down the steep slope into Mutitjulu water hole. Remarkably a number of climbers have come up from that direction. For the last 600m the path crosses bedding at right angles and you'll find yourself going up and down the valleys and crests of the surface grooves formed by differential erosion. Take care as some of these have short sections steeper than the chained section. The last time we climbed we found a number of safe short cuts were possible that provided an easier route. Shoes that provide good ankle support will help reduce the chances of ankle sprains across the uneven surface. There are many false summits on the way and you will not see the summit monument until you are about 200m from it. If it has rained recently most of the pools will be filled with water. Keep your eyes open for shield shrimp in the pools.

Figure 45: Take care negotiated the ups and downs of the deep grooves on the summit

Congratulations! You made the summit! Prior to 1986 you could now add your name to the Achievement log, but these days you can't. It's still a great spot to take a photo proving you made it to the very top.

Figure 46: Taking turns to sign the "Last Log Book" at the Summit Monument

On reaching the summit Beryl Miles wrote:[142]

Standing on top of Ayers Rock staring across at Mt. Olga lying like a heap of opalescent bubbles in the hand of the desert, the lines which were such a favourite with Giles came into my mind:

> *'Yet the charmed spell*
> *Which summons man to high discovery*
> *Is ever vocal in the outward world,*
> *Though they alone may hear it*
> *Who have hearts*
> *Responsive to its tone.'*

On the way back down the section between the monument and the chain is fairly straightforward. Just take your time. Going down the chain you may find it easier in places to hold the chain and descend backwards rather than go down face first due to the height of the chain which tends to stoop people forward as they descend due to its low height. The other way is to sit down and slide on your bum, but you may find your shorts don't survive the entire trip. Again experienced walkers will find they are able to safely walk down using the rougher rock surface adjacent the chain and careful foot placement, just remember there is less protection on the steeper sections. Take care passing other climbers.

Once at ground level be sure to look at the five memorial plaques near the base. These are around to your left if you are going down the Rock. The memorial sign for Bill Harney was once located on the sloping rock platform above these.

What can you see from the summit?

The directional marker at the summit indicates the names of the major mountain ranges visible from the summit, along with Lake Amadeus directly north. You may also be able to make out Yulara to north north east and Mutitjulu township to the east. The intervening ground is marked by a mosaic of patterns reflecting different vegetation in sand ridges and plains and dried lake beds. Lookout for distant fires, not uncommon during all times of the year.

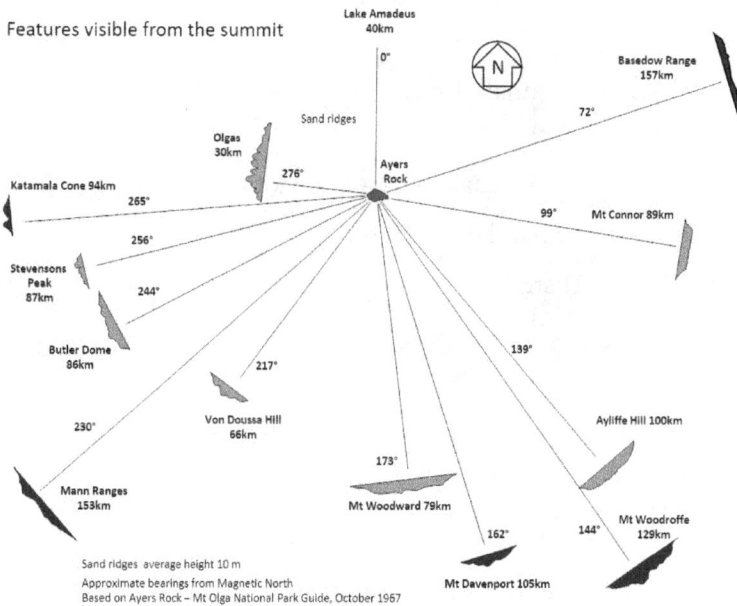

Figure 47: Features visible from the summit

Geological Units at neighbouring peaks

The geology of the main peaks is indicated below.

Olgas: Mount Currie Conglomerate (Neoproterozoic-Early Cambrian)

Katamala Cone: Dean Quartzite (Neoproterozoic)

Stevenson Peak: Dean Quartzite (Neoproterozoic)

Butler Dome: Dean Quartzite (Neoproterozoic)

Mann Ranges: Umutja Granite Suite – Various Granites (Mesoproterozoic)

Von Doussa Hill: Dean Quartzite (Neoproterozoic)

Mt Woodward (SA): Birksgate Granitic Complex (Mesoproterozoic)

Mt Davenport (SA): Birksgate Granitic Complex (Mesoproterozoic)

Mt Woodroffe (SA): Giles Complex layered mafic/ ultramafics (Mesoproterozoic)

Ayliffe Hill (SA): Shear zones of Musgrave Block – Mylonite (Palaeozoic)

Mt Connor: Winnall Beds – conglomerate and siltstone (Neoproterozoic)

Basedow Range: Winnall Beds – sandstone (Neoproterozoic)

Lake Amadeus: Salt lake with recent clay/silt (Quaternary)

(SA - South Australia)

6

Tourist Deaths Climbing Ayers Rock

Prior to July 2018, Parks Australia variously claimed *more than 30*,[143] 35[144], at least 35[145], 36[146] or 39[147] people had died climbing Ayers Rock in different departmental documents. The current "official" number is 37.[148] To clarify those fatalities, during early 2018 we requested Parks Australia provide information about the deaths it claims have occurred ON the Rock. No information was received that could substantiate the claims made, raising doubts about the numbers and about incident management at the Park. One would have thought for instance that an official register would be used to record these important details that could be called on to provide accurate information. In the absence of official assistance we based a review of the number of people who have died climbing the Rock on newspaper articles and other documents. Based on these figures we have substantiated 18 deaths to climbers ON the Rock since 1962 (See Table 1). These include six falls and 12 from natural causes, most likely related to heart failure in some way. The other deaths it seems either missed out on being reported, or were not dramatic enough to warrant media coverage – perhaps because they occurred elsewhere in the resort, sometime after people had climbed in more mundane circumstances. Linking these deaths directly to climbing in the absence of other information is not justified. It is possible the deaths were due to dehydration, pre-existing medical conditions, other activities undertaken, what was consumed at dinner or breakfast, or something else.

In an interview with Tim Webster in November 2017,[149] Park Manager Mike Misso stated: *Yeah, look over 30 people have known to have died from climbing, and what I mean by that, people could, um, you know, potentially climb it, go to the resort and then you know, could have a heart attack later.*

These comments appear to substantiate our line of reasoning that of the deaths claimed to have occurred to climbers ON the Rock by Parks Australia, twenty or so have actually occurred somewhere other than ON the Rock. We offer Parks Australia the opportunity to provide the detailed data we requested so we can update our statistics in subsequent editions. If any readers have any information that might shed some light on the actual numbers please get in touch through the publisher.

Chronicles of the fallen

We are all responsible for our decision to climb or not, and if things go wrong the ultimate blame lies with ourselves.

There are poignant lessons to be learnt from the deaths that have occurred to climbers on the Rock. Some have been due to misadventure, with some climbers wandering well off the path to places beyond their abilities. Others have had unknown medical conditions, overestimated their fitness, pushed themselves too hard, or tempted fate too often. It is clear that most of the deaths that have occurred could have been avoided.

The first five deaths ON the Rock are recorded in memorial plaques at the base of the Rock. The first tourist death occurred on 26 May 1962 when 16-year-old Brian Strieff, on a school excursion with Carey Grammar, wandered off the main path in heavy fog on the way down and fell to his death.[150] The next year, in December 1963, 25-year-old English tourist Marcia Buriston, probably affected by heat stroke, also fell and died. Newspaper reports recorded that before she fell she suddenly ran down the steep side of the rock "like a gazelle".[151] Weather records indicate the temperature was well over 100°F (38°) that day.

These deaths resulted in the chain being installed in 1964.[152] This greatly improved the safety of those on the steep climbing spur but also opened the Climb up to more people, who probably would not have thought about climbing previously. The estimated proportion of visitors climbing jumped from about 20% before to over 70%

after the chain was installed. The nature of deaths changed also with more elderly visitors succumbing to deaths from heart failure brought on by a range of factors, heat stroke, over exertion, and dehydration among them.

The next two deaths on the Rock, Leslie Twaites (15/6/1972 aged 63) and Ernest George (10/8/1977 aged 59) were due to natural causes, likely heart attacks brought on by the physical exertion of the Climb and pre-existing medical issues probably unknown to the two. The text on Leslie Twaites' memorial indicates that climbing the Rock was one of his lifelong ambitions.

Brian Miller reportedly slipped off the Rock on 18 May 1978 following heavy rain, falling over 100m to his death. Police stated that the rock was slippery after more than 40mm of rain had fallen in the past 36 hours.[153] The Ayers Rock rain gauge shows 21.9mm for the 24 hours to 9am on the 17th. The maximum temperature recorded on the 18th was 20.8°C.

British tourist David Brett, aged 31, was killed by a fall from the Rock on Australia Day 1986. He chose to climb in the late evening and would have still been on the Rock after dark. The temperature that day very close to 40°C. His disappearance was reported and police commenced searching but did not find his body until a week later on 2 February 1986.[154] He had fallen near a Dingo den and was badly mauled.[155] While searching for missing body parts, nearby police discovered a weathered white jacket with yellowed edges half buried in the sand. Such a jacket had been described in great detail by Lindy Chamberlain during her 1982 trial for the murder of her baby Azaria in August 1980. Lindy Chamberlain had insisted from the outset that a dingo had taken the baby but was convicted of Azaria's murder on 29 October 1982, and sentenced to life imprisonment. Part of the case hinged on the absence of dingo saliva on Azaria's clothing found at the scene. Lindy Chamberlain had insisted her baby was wearing a jacket that absorbed the saliva but prosecutors doubted it existed. Chamberlain identified the jacket found near Brett's body as Azaria's and was released from Darwin's

Berrimah jail just six days later. She was completely exonerated of any wrongdoing. Azaria's father Michael, who died in January 2017, had attempted for many years to place a memorial plaque to his daughter[156] on Ayers Rock, but, sadly, approaches were rejected by the Park Board.

It is difficult to find the names of the next six reported to have died on the Rock between 1987 and 1994. They were all related to likely heart attacks. In one case the deceased had had three previous heart attacks and was wearing a pace-maker; some people certainly push their luck! There is a report indicating four deaths in 1990, but no particulars are given and we are unable to confirm the location of the deaths, on the Rock or elsewhere, which were said to be to elderly, unfit climbers.[157] Without details on the location we have omitted them from our total, but if they were on the Rock, they would take the corroborated total to 21.

Two Japanese tourists died in 1996 and 1997.[158] A 32-year-old man, Hirofumi Shoji, fell 180m from the climbing spur on 20 November 1996 chasing after his backpack that had fallen down the slope. Early the following year, on 23 February, a 56-year-old man, Moriya Uchida, died from hypertension after setting out on the Climb following a large breakfast. The coroner's report indicated the heavy meal may have been a contributing factor to his death. There are reports of an Estonian man falling from the north face in 1998, but we have not been able to confirm further details. A heavy breakfast was also implicated in the death of 65-year-old German tourist Stefan Braun on 11 October 1999, who died from a heart attack on the climb up, though there were also questions concerning his overall health and fitness. Braun's death seems to have been misreported as occurring in 2000[159] and 2001.[160]

Parks Australia instituted its heavy-handed closure protocols in the early 2000s. This greatly reduced the potential for people to climb, but a few deaths have still occurred with two on the Rock this century. On Anzac Day 2010, a 54-year-old Victorian man collapsed on the way down the Climb[161] and, on 3 July 2018, a 76-year-old Japanese man[162] died on the summit plateau while walking back towards the chain. When we climbed a few weeks later, paint marks used to indicate helicopter landings in the recovery of his body were still visible.

Deaths at Kata Tjuta

For some reason deaths at Kata Tjuta are less controversial than those on the Ayers Rock Climb, yet for this century the same number have died in both places. Since 1987 there have been at least four deaths reported at various places around the conglomerate domes. Two German tourists died from heat exhaustion in 1987 and 1988. In 2002 an 82-year-old Victorian tourist died from a suspected heart attack at the Sunset viewing platform, and in January 2008,[163] a 51-year-old Japanese woman died after collapsing at the Valley of the Winds lookout, after walking 300m from the carpark.

Some lessons...

For fit and healthy people who follow the marked path the Climb is a low risk activity. Risk increases with age and with recklessness. People well into their 80s have successfully climbed without issue. It's important not to push yourself beyond your abilities. If you are determined to reach the top then just take your time. Take plenty of breaks up and down. It does not have to be a race. Be careful of having a big meal just before setting off up the chain, it may well be your last.

Comparing risks

It is difficult to find a similar experience internationally that compares to climbing Ayers Rock. It is a rather short but steep walk (1.5km) and not that high (350m) in comparison to some others around the world, but it is quite accessible meaning it attracts people with a wide range of abilities. The desert climate can be quite challenging and leads to many issues with dehydration and heat related health problems that can trigger fatal outcomes not present at other similar places.

Using Parks Australia's conservative figure of 37 deaths since the first in 1962, and given just on seven million people have climbed, this provides an average fatality rate of about five deaths per million climbers, or one death for every 189,200 climbers. The vast majority of deaths have been to people over 50 from heart attacks.

Mount Fuji in Japan is a much more arduous but much more popular climb. The death rate for climbing this sacred mountain is 15 deaths per million climbers or one death for every 64,800 climbers (2005-2013 2.4 million climbers, 37 deaths). This is three times greater than the Ayers Rock Climb and yet there are no plans to ban climbing this sacred mountain.

Only four people have died in the Uluru Kata Tjuta National Park this century, two on the Rock and two at the Olgas. This means the death rate, given approximately 6,000,000 visitors to the Park since 2000 and four deaths, is 0.22 deaths per annum, or 0.67 deaths per million visitors. In contrast, in recent years there are about 12 deaths per annum at the Grand Canyon in the USA. Since five million people visit the Grand Canyon each year this would equate to about 2.4 deaths per million visitors, a rate 3.5 times higher than that at Ayers Rock. But there are no plans to close the canyon.

The micromort is a useful concept in comparing risks of different activities. One micromort is a unit of risk defined as a one-in-a-million chance of death. If an activity exposes you to a micromort of risk it exposes you to a one in one million chance of dying from that activity. An activity with 10 micromorts represents a 1 in 100,000 chance of death.

Climbing fatalities at Ayers Rock can be divided into two main classes: climbers under 50, and those above. Of the 37 deaths Parks claim on the Rock we estimate 25% have affected people under 50 and the remainder those above. If climbers under 50 represent about 60% of the 7,000,000 people that have climbed and nine in this group have died, the exposure is about two micromorts per climb, about the same as driving a car 800km. The risks are greater for those over 50. If 40% of climbers are over 50 and 28 have died, the increased exposure would be about 10 micromorts per climb.

For comparison driving the 2,800km from Sydney to Ayers Rock provides seven micromorts of increased risk (400km of driving = 1 micromort).[164] If you rode a motorcycle for the same trip you would be exposed to 280 micromorts (10km of motorbike riding = 1 micromort).[165] Skydiving will cost you eight micromorts per jump, while scuba diving between five and 10 micromorts per dive.[166]

Overall the Climb is a low risk activity for fit and healthy people. If you are over 50 and you drove to the Rock from Perth, via Adelaide the Climb has about as much risk as the drive there.

Table 1: List of corroborated deaths ON Ayers Rock

Date	Name	Age	Reason	Source
26/05/1962	Brian Streiff	16	Fell after moving off the main climbing path in mist	Plaque
22/12/1963	Marcia Buriston	25	Fell during climb, heat stroke?	Plaque
15/06/1972	Leslie Twaites	63	Natural causes - likely heart attack	Plaque
10/08/1977	Ernest George	59	Natural causes - likely heart attack	Plaque
18/05/1978	Brian Miller	25	Fall - slippery after rain	Plaque
26/01/1986	David Brett	31	Fall - led to discovery of Azaria Chamberlain's matinee jacket and release of Lindy Chamberlain	*SMH*, 5 February 1986
1/06/1987	Name not known	72	72-year-old with diabetes, had 3 previous heart attacks and fitted with pacemaker	*SMH*, 10 September 1987
11/02/1991	Name not known	60s	British man in his 60s collapsed while descending.	*Canberra Times*, 13 February 1991, 22
August 1991	Name not known	44	Myocardial Infarction: 10:30 Male, Canada. Base of climb	Salib and Brimacombe 1994. *Medical Journal of Australia* V161, 693-694

October 1991	Name not known	53	Myocardial Infarction: 07:00 Male, Canada. Base of climb	Salib and Brimacombe 1994. *Medical Journal of Australia* V161, 693-695
October 1991	Name not known	62	Myocardial Infarction: 08:30 Male, Canada. Base of climb	Salib and Brimacombe 1994. *Medical Journal of Australia* V161, 693-696
1/07/1994	Name not known	52	Likely Heart Failure, turning back from climb	*SMH*, 23 July 1994, "The rock claims 26th victim"
20/11/1996	Hirofumi Shoji	32	Fall chasing backpack (Japanese Tourist)	Dead in the Heart *The Bulletin*, 16 November 1999
23/02/1997	Moriya Uchida	56	Heart failure Natural causes (Japanese Tourist)	Dead in the Heart, *The Bulletin*, 16 November 1999
1998	Name not known		Young Eastonian man climbing in illegal area fell from north face	*SMH*, 29 December 1998
11/10/1999	Stefan Braun	65	Heart Attack top of chain (German tourist)	Dead in the Heart, *The Bulletin*, 16 November 1999
27/04/2010	Name not known	54	Heart attack Base of climb after turning back from chain	http://www.news.com.au/travel/travel-updates/uluru-death-fuels-call-to-close-rock-climb/news-story/2b81fa8bce0b63094f6e45838f091c3b
3/07/2018	Name not known	76	Heart Attack, on summit on way back down. Japanese tourist.	http://www.abc.net.au/news/2018-07-04/japanese-tourist-dies-climbing-uluru/9937848

7

AYERS ROCK AND THE 20% MYTH

In its *2010 Management Plan* Parks Australia outlined a series of measures it would use to justify a ban on climbing Ayers Rock.

From section 6.3.3 (C):

(c) The Climb will be permanently closed when:

- the Board, in consultation with the tourism industry, is satisfied that adequate new visitor experiences have been successfully established, or

- the proportion of visitors climbing falls below 20 per cent, or

- the cultural and natural experiences on offer are the critical factors when visitors make their decision to visit the park.

None of the conditions outlined above have been met. Tourists have been omitted entirely from the consultation process, and any research on visitor intentions, particularly surveys, has been conducted by groups prejudiced against the Climb. The surveys are heavily biased (in one study there was a bias of almost 2:1 in responses from females over males) and they have no statistical merit. Alternate visitor experiences, such as Segways around the rock, bush BBQs and the coming *Uluru Skyship* stationed 9km from the Rock that only ascends to half its height, are in no way comparable to the exhilaration, wonder and joy imparted by the Climb. All these extras come at an additional substantial cost to the Park entry fee.

In regard to the second condition, that the proportion of visitors climbing falls below 20% Parks Australia installed pedestrian counters on the Climb and recorded data from June 2011 to the end of June 2015 to establish the proportion of people climbing.

I obtained the results of this data and an analysis through a FOI request. Re-analysis of the data does not support Parks' contention that the 20% figure has been met when the availability of the Climb is taken into account. Instead, it shows that on average for days tourists are able to make a fair choice between the Climb and other activities – an average of 44% choose to climb, with the highest figures well above 70%. The FOI documents also reveal many problems were experienced in obtaining reliable data, with counters being down for much of the assessment period, counters under reporting the number of climbers by at least 30%, calibration issues with counters, and problems calculating total visitor numbers (kids are missing from the 2011 data). It's a mess that does not stand up to close scrutiny!

For the period the counters were in place (between 1/6/2011 to 30/6/2015) Parks Australia's onerous closure protocols meant the Climb was fully open a mere 10% of the time (151 days out of 1491). A fair estimation of tourist preferences can only be made for those days that a full range of choices are available from when the Park opens. Tourists visiting Ayers Rock are typically under tight time schedules and if a given activity is not available they quickly move on to other options. There is no time to wait at the base of the Rock for the Climb to open. Most tourists will undertake only one strenuous activity, either one of the walks at the base, the circuit walk, or the Climb; and then head back to the resort to relax.

Under reporting

A significant issue flagged in a report analysing the data for Parks Australia (Becken 2017) is the under reporting of climbers by counters. Feedback from experts suggested a figure of 30% was a fair estimation of the amount of climbers not recorded. This occurs as the counters record groups of climbers, rather than individuals. Counters also recorded data on days the Climb was officially closed suggesting illegal climbs, saying a lot about visitor intentions, with sometimes recorded numbers higher than the number of daily visitors. These problems were discussed in internal memos but apparently never fully rectified and the counters were withdrawn from use in June 2015.

Findings of Parks Statistical Analysis

Parks Australia commissioned Prof. Susanne Becken (Professor of Sustainable tourism) of Griffith University to undertake a statistical analysis of climbing data from June 2011 to the end of June 2015. Prof Becken's views on the Climb can be found in an article she wrote for *The Conversation* titled "Closing Uluru to climbers is better for tourism in the long run". In the article Becken and colleague Michelle Whitford write: *Closing Uluru for climbing should be seen as a shining example of sustainable tourism being a vehicle for the preservation, maintenance and ongoing development of culture, traditions and knowledge.*

We leave questions of the suitability of Prof. Becken to undertake an unbiased study of the climbing data to readers.

Prof Becken concludes her analysis addressing the 20% question:

To address the main question of whether the proportion of climbers is below or above 20 percent the answer depends on the method chosen (as illustrated in this report) and also whether particular months are of interest or an average over a year is taken.

In other words, if you are fair and take the availability of the Climb into account, the 20% cut off has not been reached. If you consider the entire time period including all those days the Climb is completely closed (49% of the time!) then less than 20% are climbing.

A figure from her report (see Figure 48) shows a wide range in the proportion of climbers on a daily basis. Most notable is the seasonal trend reflecting the effective complete closure of the climb between October and April, during warmer months, and the gap in data in 2013-2014 associated with the counters not operating. These un-adjusted results, including partly open days, show many occasions where the proportion climbing exceeds 20%. In her report Becken did not separately assess the proportion of climbers on fully open days from those days the Climb was only partly open. Partly open days, included in Figure 48 below include days the Climb was only open for an hour. In these circumstances it is not valid to use these figures to determine the intentions of tourists.

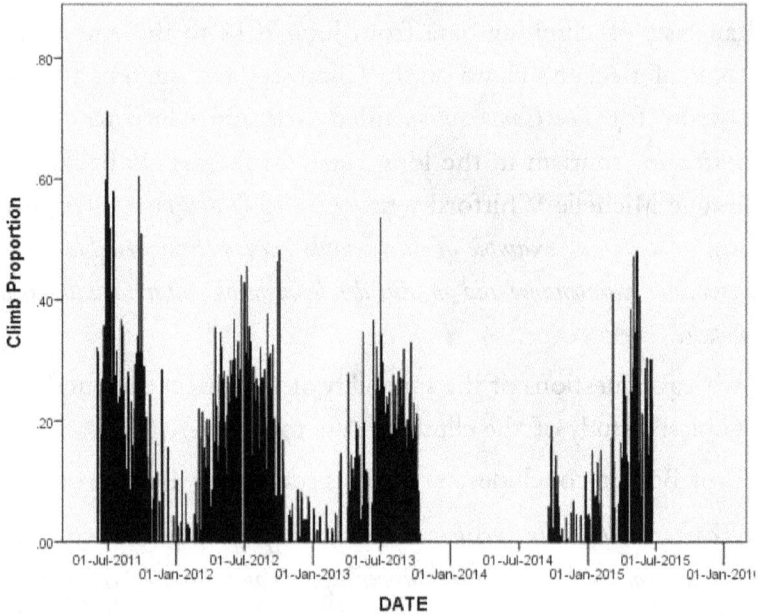

Figure 48: From Becken, 2017. Proportion of visitors climbing Uluru relative to all visitors on days the climb was fully and partly open

Re-analysis

Parks data was re-analysed looking at the proportion of tourists climbing on those days the Climb is fully open and data is reliable. Only on these days can the intentions of tourists be fully assessed. Data was analysed for:

1. all days the Climb was fully open regardless of the data reliability,

2. all days the Climb was fully open for days with reliable data, and

3. all days the Climb was fully open for days with reliable data, adjusted by 30% to account for under-counting.

Results

1. The average proportion of those climbing for days the Climb was fully open regardless of data quality (includes many days of missing counts and days of obvious significant under reporting) is 26.8% (>20%).

2. For days the Climb was fully open with reliable data: 34% (>1.5x20%).

3. For days the Climb was fully open with reliable data, adjusted upwards by 30% recommended by Parks Australia: 44% (>2x20%).

Ayers Rock % Climbers (Days Fully Open)

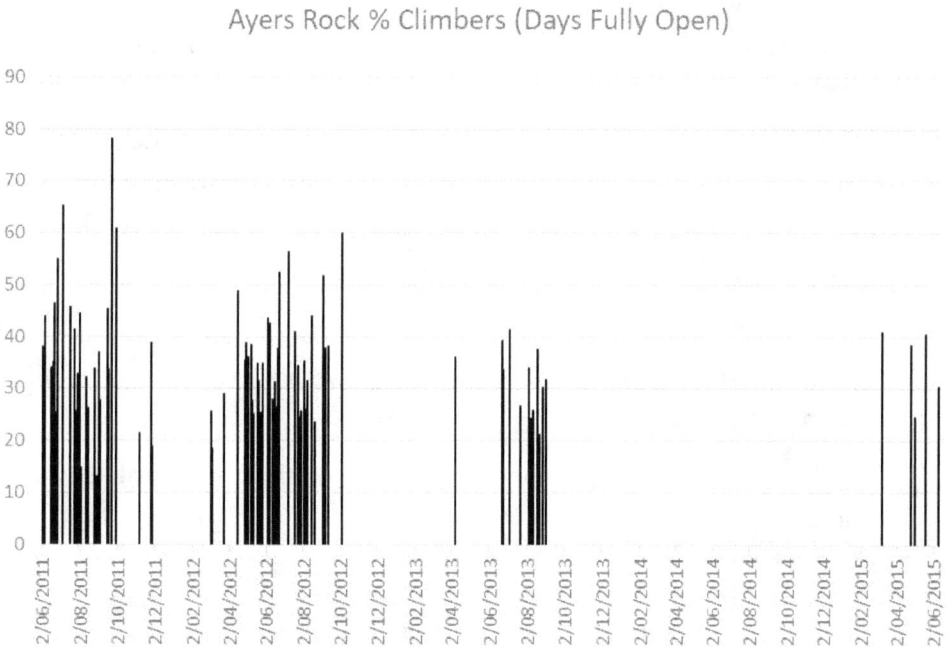

Figure 49: Average proportion of climbers on days tourists are offered a full choice of options (un-adjusted) average is 34%

The data also show no substantive change in the average proportion of climbers over the survey period:

Annual averages for years with data (unadjusted):

2011: 36%

2012: 34%

2013: 29%

2014: Insufficient data

2015: 35%

Conclusions

When visitors are offered a full range of options from the moment the Park is open, the average proportion deciding to climb Ayers Rock remains well above the 20% cut off mark determined by Parks Australia, and is likely as high as 44%, on some days exceeding 70%. There has also not been a substantive change in the proportion of climbers over the period covered by the monitoring data. The notion that less than 20% of visitors want to climb is complete nonsense.

Criteria used by Parks Australia to close the Climb have not been met and there remains no justification for closing this natural wonder. The decision to climb, or not, remains best left to individuals.

References

Becken, 2017. *Analysis of Uluru visitor climb data and monitoring methods.* Griffith University. Obtained under FOI.

Parks Australia: Climbing data and associated information obtained under FOI.

8

A CASE AGAINST THE BAN

The 1,334 km² of land that forms the National Park was signed over to the Uluru-Kata Tjuta Aboriginal Land Trust on 26 October 1985. As private property the owners are entitled to set the rules for visitors they permit on their land. If they choose to lock the gate then private citizens have no other legitimate recourse but to stop and turn around. But, and it's a big BUT, the owners agreed to lease their land back to Australians to be managed as a National Park. While the lease is in place the Park Managers, the Board and the Administrators are obligated to observe the conditions of the lease and other laws that govern access in a National Park. National Parks are shared spaces for the whole world to enjoy – free from religious and political interference, regardless of who owns the land they sit on.

Banning the Climb breaches the lease agreement

The 1994 lease agreement between the Uluru-Kata Tjuta Aboriginal Land Trust and the Director of Parks Australia, section 17 (2) states: *The lease covenants that the flora, fauna, cultural heritage, and natural environment of the Park shall be preserved, managed and maintained according to the best comparable management practices for National Parks anywhere in the world or where no comparable management practices exist, to the highest standards practicable.*

As this book demonstrates, the act of climbing Ayers Rock and enjoying the views from the summit are important items of Anangu and non-Anangu Cultural Heritage significance. The chain, summit monument and the five memorial plaques are important items of historical physical heritage associated with the Climb. The Climb has been subject to and inspired elements of the Anangu Tjukurpa along with non-Anangu literature, works of art, photography and is ingrained in the Australian and international psyche. Climbing the Rock is unarguably an important cultural heritage practice for visitors and custodians alike and as such according to section 17(2) of the lease it *shall be preserved, managed and maintained according to the best comparable management practices for National Parks anywhere in the world.* It is clear by their actions that the Board and Parks Australia have failed to recognise the cultural heritage significance of the Climb for some time. The negative propaganda about the Climb needs to be redressed and the next management plan must take the cultural heritage significance of the Climb into account, and preserve it. The Climb needs to be kept open, otherwise it seems the lease is irrevocably breached.

If the ban proceeds the lease is essentially null and void, the Director of Parks Australia must withdraw from it, and the land owners can run their land without the support of the Australian government and taxpayers, as a private park, a cattle station or whatever they want and grant access according to their whims like any other owner of private property.

Banning the Climb breaches Discrimination Acts

Under the current lease it is not possible to ban the Owners from accessing the summit via the current climbing route, or other means. Clauses in the lease agreement (Section 2) give them special rights to go anywhere in the Park they want. If the ban on accessing the summit by anyone else proceeds the following breaches of Australian Law, that Parks Australia is obligated to observe, are identified:

- The ban on climbing discriminates against non-Anangu persons participating in the climb and as such breaches the Racial Discrimination Act 1975.

- It effectively discriminates against all children participating in the Climb and as such breaches the Age Discrimination Act 2004.

- It effectively discriminates against all women participating in the Climb and as such breaches the Sex Discrimination Act 1984.

While the National Park exists, if Anangu are able to climb for cultural reasons for *inma* ceremonies or other purposes, and visitors are not allowed to climb to exercise their own equally important and legitimate cultural heritage, then non-Anangu visitors are clearly being discriminated against. Climbing the Rock is a long lasting practice of Australian, European, Japanese, Chinese, Indian, African and American cultural heritage significance. Unless the ban also includes the Anangu, it seems it would breach Australian anti-discrimination laws to which Parks Australia is obligated.

In closing your honour

In its 1985 management plan for the Park the National Parks and Wildlife Service stated that the purpose of the National Park included special outdoor recreation opportunities. The Climb is such an activity that is of special cultural heritage significance and is protected as a cultural heritage item under clause 17(2) of the current lease agreement. The Climb has brought exhilaration, joy and wonder to millions of visitors. This is acknowledged in past marketing material produced by Parks Australia and the Anangu Traditional Owners and Mutitjulu Community. Banning the Climb is disallowed by section 17(2) of the lease.

Against the provisions of the lease agreement, Parks Australia and the Park Board have discouraged millions from undertaking an activity that would happily be promoted anywhere else in Australia and the rest of the world. This is reprehensible and is reflected in lower visitation numbers since the year 2000. The loss to the local economy is significant and estimated to be at least $70,000,000 annually.[168] The next management plan must redress this injustice and take the cultural heritage significance of the Climb into account. The Climb is something to celebrate.

Parks Australia and the Board have never presented positive aspects of the Climb, and its importance to visitors and to the Traditional Owners, and in doing so have not provided them with sufficient information to make an informed decision about the future of the Climb. This defies the philosophy under which the park was established. This took a multitude of views into account in managing the Park, including those of tourists interested in climbing the Rock.

The Board went against the express wishes of former Prime Minister Kevin Rudd, and then Opposition leader Turnbull, in including closure provisions in the current management plan without adequate consultation, regarding the impacts of the plan, with the Australian people. The overwhelming feeling following announcements by politicians in 2009/2010 and media coverage at the time was that the proposed ban would never be enacted. This was reaffirmed by the Turnbull Government in 2016 which stated there were *no plans to change current arrangements*.[169] Despite the overwhelming bipartisan political support for the Climb and the overwhelming support of the Australian public the legislation to trigger the closure somehow still passed parliament.

Like it or not, climbing is the activity that has made the Park famous, and the activity deserves to continue regardless of the numbers who are interested. People in a liberal democracy should be free to enjoy the natural world, unfettered by transitory ideologies, strange religious beliefs and political views. Moreover, there is little risk (less than two micromorts to fit and healthy individuals), and the environmental consequences are negligible and can be managed in a manner to make the activity sustainable. Climbing the Rock is an innocent, low risk exercise that is part of our collective human experience. It celebrates our collective human achievements of discovery, scientific advancement and understanding of the natural world. It is likely that people have been climbing Ayers Rock since humans first arrived in the area about 35,000 years ago. Climbing, or hiking up natural features, is a worthwhile activity universally undertaken by all cultures and encouraged at many other similar locations in Australia and internationally. Only in Australia, at Ayers Rock, are people made to feel guilty by Parks Australia and the Board for contemplating, or undertaking it. This must cease.

While the beliefs of the Traditional Custodians deserve respect, the notion that their beliefs are to be forced onto others who do not believe them, in the context of simply ascending a natural feature that has existed for millions and millions of years before those Custodians arrived, smacks of intolerance and discrimination. It may even breach section 116 of the Australian Constitution that forbids the Commonwealth from making laws imposing religious observance. If local Aboriginals choose to restrict access to the summit to a few older initiated men on the grounds of their religious beliefs that is their decision. They rob no-one but their own community, their own women and children, of the joy and sense of wonder of the Climb, of the inspiration it provides and the remarkable views you can see at the summit. The ban must not be imposed on visitors who do not share those beliefs; and for whom climbing is an important, integral part of their own culture and cultural heritage.

Parks Australia's maintenance of the National Park as a cultural museum risks damaging the long term well-being of the Anangu people, slowing their assimilation into the modern world, which will inevitably happen through education and exposure to new ideas. The unchanging Tjukurpa will bend and break as other outdated ideologies have done in the past.

Regardless of the legality of the climbing ban and breaches of the lease agreement, visitors should be able to continue deciding for themselves whether or not to climb the Rock. Allowing people to decide for themselves is supposedly a feature of Tjukurpa and in line with concepts of liberal democracy under which modern Australia was founded. Imposing restrictions on a walk that has been accessed by millions of visitors over the past 35,000 years goes against reason and is an affront to humanity.

In light of the evidence presented herein, banning the Climb is simply not justifiable on any grounds as it breaks the lease agreement and unfairly discriminates against tourists who simply want to enjoy the natural world.

Parks Australia and the Board must remove the ban, restore freedom of access to the Climb and open up areas that were once freely accessible, so all visitors can experience the joy and wonder of exploring this remarkable place.

Other peaks

The ban on climbing Ayers Rock will no doubt galvanise and accelerate demands from other indigenous groups to close off other natural wonders in Australia from the general public and international visitors. What a welcome to country! As I write Mount Warning in northern New South Wales (NSW) and St Mary's Peak in the Flinders Ranges in South Australia are facing restrictions on access to their summits.

Mt Warning

The summit of Mount Warning, at 1156m, is the first place in eastern Australia to see the sunrise each day. It provides remarkable views over the surrounding Cenozoic volcanic caldera complex, forest, coastal plains and ocean beaches. Access to the summit is via a well formed walking track and the upper section includes a steep rock slope where climbers are assisted upwards by a chain. There is an established lookout at the top. The 8.8km return walk from the car park to the summit takes about five hours according the NSW National Parks and Wildlife Service (NSW NPWS). At present NSW NPWS provide the following information about the walk on their website:[170] *Wollumbin (Mount Warning) summit track is a sacred place to the Bundjalung People, and was declared an Aboriginal Place in 2015. Visitors are asked to respect the wishes of the Bundjalung Elders and avoid climbing this very difficult track.*

No space is provided for the cultural views of the over 100,000 people who choose to climb to the summit annually. Our request to have the following paragraph added to the web page was rejected:

Walking and exploring natural places is an important Australian and international cultural tradition that deserves to be acknowledged, cel-

ebrated and respected. Walkers should not feel guilty about exercising their own cultural heritage and enjoying the natural world.

Asked about the prospect of a ban the NSW government was not prepared to provide assurances the Climb to the top of Mt Warning would not be stopped. Incredibly NSW NPWS, an agency funded by all NSW residents to look after our National Parks for ALL of us, actually supports closing the walk: *The local Aboriginal Elders requested that people not climb the summit due to its cultural significance. The NPWS support this request through signage on the Summit Walking Track.*

The views are among the best from any summit on the east coast, and the walk itself is a real adventure, the reason so many make the Climb each year. To our great sadness, in their signage NSW NPWS do not mention anything positive about the summit walk, or encourage visitors to complete this Climb. For NSW NPWS it seems the walk is an inconvenience they would rather do without. All too hard to manage, it is simpler to shut it down, the reason does not really matter.

St Mary's Peak

At 1171m, St Mary's Peak is the highest point in the Flinders Ranges in South Australia (SA). It's the mid point for a spectacular scenic 20km bushwalk that rewards with 360° views of the rocky ranges, salt lakes and amazing desert plains. The Flinders Ranges provide among the best exposures of Late Proterozoic and Palaeozoic sedimentary rocks anywhere in the world and this extraordinary geology is laid for visitors to view along the walking trails and from the summit. There is an existing marked trail all the way and the footsteps of walkers do no environmental damage

This wonderful challenging walk, accessible to most people of moderate fitness with a little planning and a willing spirit, has been under threat for some time by the beliefs of the local Adnyamathanha people who believe that the peak is the head of an ancient serpent whose petrified body forms the walls of Wilpena Pound and they prefer you don't visit it. Instead they and the SA Parks Administrators want you to stop at Tanderra saddle (Alt 947m) 224m lower down where the views are less inspiring. If the Adnyamathanha don't want to visit the summit that is their decision. Like the Anangu at Ayers Rock they rob no one but themselves of the joy and sense of wonder available at the top. Other visitors who do not share these views should be able to decide for themselves if they want to climb the whole way, and they should be allowed to do so without being made to feel guilty about enjoying and learning about the natural world.

In January 2018 I wrote to the SA Government asking them to provide guarantees that the walk would remain open, but their spokesperson was unable to give any guarantee the walk would not be banned.

Access at risk

The language being used about access to both these wonderful natural spaces is very similar to that used at Ayers Rock. The 2010 Uluru-Kata Tjuta Park management plan provided for the Ayers Rock climb to be closed. This plan slipped through parliament despite bipartisan political support and the backing of the public for the Climb. In 2016 the current Federal government gave assurances that the Ayers Rock Climb would not be closed, and yet somehow myth and superstition have won over logic and common sense, and barring a miracle we will likely see the Climb closed in October 2019.

With both the NSW and SA governments doing nothing to prevent a ban at both their peaks, and with both State Parks departments openly supporting a closure and not encouraging climbing, there remains a very real chance that the walks to the summits of Mt Warning and St Mary's Peak will end up the same way as the Ayers Rock Climb. All it will take is for one soft headed bureaucrat to quietly write the paragraph and insert it as an amendment to the current management plans, thereby severing another beautiful part of the country from public access.

Freedoms we once thought were immutable are under threat

from forces prepared to tear down our Western civilisation and unleash a new dark age. Your job should you chose to accept it is to actively support access to all our natural wonders for all visitors regardless of race or religion; and to keep an eye open for any proposed amendments to the management plans of these areas that could make a ban a depressing reality. Make sure your local politicians receive a clear message that these wonderful natural places belong to all of us, and should remain open to everyone, not just a select few. Let them know you will vote against them, if they vote to enact a ban or restrict access to our collective natural inheritance.

9

A Climbing Tale

Everyone who has climbed the Rock has a story to tell, this is mine.

First climb 1998

In 1998 I started a job as a regional mapping geologist for the Northern Territory Geological Survey. My work required travel to some of the remotest regions in central Australia including parts of the Tanami Desert only accessible with ease in a helicopter. Coming from the lush green forests of the east coast the desolated red centre makes for a stunning contrast, a visual feast that leaves a lasting impression. The impact of first glimpsing Ayers Rock and The Olgas standing prominent above the flat sand plains is especially memorable. I first saw this great red rock driving down the Lasseter Highway from Alice Springs to meet Matt Golombek, then Science Director of the Mars Pathfinder Mission shortly after starting with the survey in Alice Springs. Matt was on a tour of central Australia as part of a series of educational lectures on the epic Mars Pathfinder Mission. It seems central Australia and Mars have a few things in common. My job along with a colleague, was to act as Matt's tour guide on a whirlwind geological tour of the Rock, The Olgas, Kings Canyon, Gosses Bluff (a spectacular meteor impact crater that deserves more visitor attention) and the eons old West MacDonnell Ranges.

We left the Alice in a heavily loaded 80 series landcruiser wagon. Coasting down the Stuart Highway at about 150km/h per hour (this was prior to the speed limit being introduced) the needle on the fuel tank perceptively moving as our speed and the heavy bull bar and loaded roof racks provided a drag on our progress south. We stopped briefly at the Erldunda roadhouse, then turned right on to the Lasseter Highway. The flat sand plains, broken by slight rises over the sand ridges, are unrelenting. About 14km east of the airport turnoff looking west you get your first glimpse of the Olgas as a set of rounded red hills on the western horizon and looking left at about the same point you get your first sight of Ayers Rock, peeking above the dunes. The dunes keep it largely hidden all the way into Yulara. The first and best uninterrupted views from the road occur at the sunset viewing area about 4km from it. Baldwin Spencer took his photo from near this point in 1894.

After collecting Matt at the Ayers Rock airport, "fresh" off the plane from Los Angeles via Sydney (a mere 20 hours or so of flying time) we almost literally threw him in the back seat of the Toyota and headed out for a four hour trek around the Olgas Valley of the Winds walk. These conglomerate domes are arguably even more impressive than Ayers Rock, especially due to the variety of lithologies and shape and sizes of the rounded boulders that make them up. We stayed overnight at the Yulara Resort taking advantage of the BBQs to cook a steak for dinner. Early the next day we drove down to the base of the Climb and headed up. The day was quite warm, in the high 20s by the time we started climbing. The three of us enjoyed the camaraderie and physical exertion of the Climb and soaked in the glorious scenery along the way with hundreds of other tourists ranging in age from under seven to over seventy. The views along the way, and at the top were extraordinary, life affirming and humbling. From a geological perspective, the summit provides a point from which to ponder the hundreds of millions of years that lie behind geological and geomorphological wonders laid out before you. We took the obligatory photos and headed back down. We did not have time for the base walk and continued on to Kings Canyon arriving there before lunch to walk the rim. After this we pushed on, stopping to camp overnight, laying out swags under the stars and cooking dinner around a fire just off the Mereenie Loop Road in sight of Gosses Bluff. We would return to Alice the next day. I promised myself I would be back to take it all in again at a slower pace.

My wife and I climbed in early 1999. We also did the circuit from Alice Springs, stopping first at Ayers Rock, The Olgas, then Kings Canyon, and back via the loop road through the Western MacDonnell Ranges. On top we spotted a huge perentie that hissed at us and scuttled away on our approach. We left Alice for Sydney in 2001 and I always knew I'd be back to the Rock to show off that view to my children... one day.

Figure 50: Perentie on the Climb, 1999

Climb for science, 2018

Without the prospect of a ban on climbing I probably would have waited a few more years and done it as part of a longer road trip around the country. The cost of getting there and staying at Yulara is ridiculously high so if you can add other attractions and experiences like Alice Springs and MacDonnell Ranges all the better for value for money. But the decision to ban the Climb in 2019 changed that and necessitated a different approach if my children were to experience the same feelings of wonder and awe I and millions of other climbers have experienced. Who knows? They could shut the whole thing down tomorrow. So another mad rush to the Rock took place.

Looking at the closure statistics, winter is easily the best season for climbing, when the gate is mostly open, so I booked flights and accommodation to have us at the Rock in time for the anniversary of the first recorded climb by William Gosse, on 20 July. For the period 2011-2014, the Climb was fully open or partly open in June 66% of the time, July 74% of the time and August 78%. August generally has the highest number of fully open days. Winter also has better weather and there are fewer flies this time of year. Rather than being a slow burn, this was now a fly-in, fly-out operation. Our visit would also celebrate Gosse and the achievements of other explorers and scientists in central Australia. We invited Parks Australia to join in our *Climb for Science* but they declined writing: *Parks Australia will not promote the climb for science, as we support the wishes of the Anangu traditional owners who request people do not climb Uluru.* Sad that in this modern age our Parks Administrators choose superstition over science!

It's hard for tourists who have not previously visited the area to appreciate just how much of the Ayers Rock experience has been lost. The highly regulated nature of the Park these days is in stark contrast to what it was like in the early 1980s when access was much less constrained and people were not made to feel guilty for simply enjoying the natural world. With some irony the Traditional Owners at the time supported the open door policy and had only closed off access to the Men's Initiation cave and the Ngaltawata Pole. They were happy for tourists to go anywhere else.

The changes I found since 1999 were quite profound. The fence installed at the bottom of the Climb and the gate with its ridiculous pronouncements about why the Climb is closed are an insult to the memory of Tiger Tjalkalyirri, the Rock's first indigenous tourist guide who climbed many times. The protocols that keep the Climb closed 80% of the time are needlessly conservative and arbitrarily enforced. We noticed the number of regulatory signs on parking, access, photography, eating and drinking have increased substantially. As Ross Barnett wrote in *4 x 4 Australia* in 2010 the place has gone from Ayers Rock to UluRules.[171] The ever present message is that Big Brother is watching. Don't dare enjoy yourself here as we have many ways to punish you. The signs around the base walk, particularly the northern section, succeed only in spoiling the views. The number of photos of various sacred sites available on line make a mockery of the heavy-handed warnings against photography. It was sad to see access at the Olgas has been further reduced. The wonderful views from the Kata Tjuta Lookout are no longer available, making it impossible for me not to ask why this should be the case. What does anyone achieve from limiting opportunities for wonder and awe? The costs for visitors in terms of what has been taken away are immense. However, unless they had visited prior to 2000 they probably won't know how much they are missing.

We arrived by plane from Sydney at around 3pm and after waiting for a German family to argue with the hire car firm on the suitability of 2WD vehicle on the Mareenie Loop Road (Red Centre Way) we were off and running. I had pre-paid for the Park tickets and we were quickly through the gate. From a short distance we could make out some small dots on the climbing spur. It seems the Minga are active today! We quickly found a carpark at the base, tossed some water bottles in the bag pack and off we went before the fickle rangers could change their minds. We ignored the message signs at the base as we were climbing to celebrate our own culture and we had permission to climb from past elders Tiger Tjalkalyirri, Paddy Uluru and Toby Naninga. Tjukurpa is unchanging and these men were more closely connected to their land and more knowledgeable of their laws and customs than those sitting on the current Board. The girls didn't need to be asked if they wanted to climb and were quickly up past chicken rock to the bottom of the chain before I had time to get a camera out. A young lady on hands and knees was rather stunned at our rapid progress upwards: *How did you do that?* The path along the chain is a little smoother than I remember. I've been told it's a scar on the Rock. In reality it's a strip only a metre wide that is slightly smoother than the rest of the Rock. Looking down the path it is a little more reflective with the sun in the west than the adjacent rough surface. From the sunset viewing area 4km away it is not noticeable, unless you know where to look. Another myth falls by the wayside. As on previous climbs, I choose to ignore the chain and walk up unassisted, mindful of Arthur Groom's words that the Climb here is *nothing else but a strenuous and spectacular uphill walk.* The rough surface adjacent the worn strip below the chain provides ample grip, even to my well-worn scarpas. You can walk up even the steepest sections by carefully placing your feet at the scalloped edges of surface pits and hollows. The girls stuck close to the chain and took their time, with the narrow section nearer the top of the chain making them holding on tightly. I took out my 1967 guide book and looked for the Katamala cone south of The Olgas. As we climbed, it slowly took shape, as did many other mountain ranges on the southern horizon.

I was surprised at how many people were climbing. The Board would have you believe that less than 20% of visitors want to climb, but this afternoon it seemed like this was the place to be. Had the Climb counters been operating I suspect the numbers today would be approaching past levels of about 70%. At the top of the chain we took a short break at the flat landing area that has good views of the Olgas.

Bromley Climbs Again

I had brought Bromley with me and took a photo of him with the Olgas in the background. In 1986 Alan and Patricia Campbell chronicled and photographed the amazing adventures of Bromley Bear in central Australia. The Campbell's published this captivating children's tale in 1993 under the title: *Bromley Climbs Uluru*. After some desert fun with Skye the Unicorn and his best pal Koala, Bromley climbs the Rock. On top he meets a dingo called Kurpanga who shows him a waterhole and a safe way back down.

According to the Campbells, *Bromley, the outdoor adventure bear is the Indiana Jones of the bear world. Bromley is more than just a bear with a name, he is the outdoor bear with attitude. He does not sit back and dream of adventure, he is out there experiencing the danger, the joy, the fun of living an adventurous life. Bromley's Motto is: "Let's do it."*

What a wonderful message to pass on to the next generation!

The book was sold at the bookshop at Yulara for nine years! About 40,000 copies were sold around the world. In 2003 the Campbells were threatened with a $50,000 fine and told to rewrite the book by the Central Land Council and National Parks because the Traditional Owners found the book offensive. Draconian regulations hidden deep in the Act that governs the Park provide Parks Australia with power that would make Big Brother blush. Parks wanted the book rewritten and retitled *Bromley Visits Uluru and has a Cultural Experience*. Seriously! This an example at how overly sensitive and sanctimonious Park Management and the Board have become. Fortunately the Environment Minister at the time, Dr David Kemp, did not pursue the case. Kemp said, in his opinion, court action against the Bromley authors was *not appropriate given the importance of principles of freedom of expression in our society*. We hope the same still holds for this book!

The path from the top of the chain to the summit is marked by a white painted dashed line. In contrast to my last visit, this time all the water holes were dry.

Figure 51: Bromley Climbs Again!

The girls enjoyed traipsing over the undulating summit. We stuck roughly to the marked path. Like the chain section the rock is worn a little along the white line and you can find better grip by walking a little alongside it. A couple of short cuts here and there and soon you are at the summit monument. A small cluster of climbers stand around taking photos with the sun setting in the west.

When we climbed in the late 1990s the bronze plaque at the summit monument was missing the map of Australia. Now the Australian coat of arms had also disappeared. The shelf that once housed the summit logbooks had been clumsily sealed over, an insult to the care and good workmanship of NT Parks Staff in erecting the monument in 1970.

The Last Logbook

I had with me a plastic container holding a small exercise book, a copy of the more practical 1967 Park Guide and a copy of the front cover of the 1981 Guide that featured the Climb prominently. The exercise book was titled *The Last Logbook*. The front cover read:

Signing the summit logbook has been an important cultural institution at Ayers Rock since the 1890s. Sadly, since the late 1980s Park Management have denied Australians and International visitors the opportunity to record their achievement.

The first climbers to leave a note marking their achievement were Allan Breadon and W Oliver on March 4, 1897: "We added a few stones to the pile and left two wax vesta boxes (tins) with names and date thereon."

Glass coffee jars held the names of climbers between 1932 and the 1950s. In September 1950 the jars held the names of about 70 climbers.

Formal log books, termed the "Achievers' book", replaced the assorted collection of jars and tins lodged at the summit cairn in 1966. These were maintained by the Conservation Commission of the NT into the late 1980s, until Parks Australia without consultation, and against the conditions of the lease agreement, stopped maintaining them. Shamefully in the 2000s the log book shelf in the pedestal was clumsily sealed over preventing access – an act of vandalism against our cultural heritage.

For the period between 20 May 1966 and 24 May 1986, 171 logbooks recorded the names of about 1.3 MILLION climbers. Until the pedestal was completed in 1970 log books were housed in a tin container on the stone cairn. The whereabouts of the jars left by McKinnon with the slips of paper signed by the first climbers is unknown.

This is the LAST logbook on Ayers Rock. We request the last person to sign it, place it in the envelope provided and post it. We will direct it to join the 171 others in the NT Archives.

The girls and I signed it and placed it back in the plastic box at the base of the monument. A rock on top preventing the winds from blowing it away. My intention is to collect it on our last day of our visit, but just in case I have also left a stamped and addressed envelope if we don't make it back up for some reason. We take a last photo for climb for science, myself celebrating at the summit with a

Figure 52: At the summit with the last logbook. Note the state of the summit memorial July 2018. Both the Map of Australia and Australian Coat of Arms missing (Photos Zoe Hendrickx)

double Vulcan salute, in awe of the achievements of past investigators. By standing on their shoulders we can see a little further over the horizon. We head back down taking a few short cuts to the top of the chain. The girls carefully make their way back down. When we reach the bottom the gates have been closed.

The next day we cycle around the Rock heading off from the Cultural Centre carpark to the start of the Mala walk. The Climb is open again and beckons. We can see as we ride past there are at least 100 people at various points on the chain, once again making a complete mockery of Parks' 20% claim. A few tourist groups are herded like cattle along the Mala Walk being filled with standardised messages from seemingly bored guides, none of whom appear to be locals. It's a pity Tiger is not around to show them the Putcha dance! We duck into a few of the caves. For me the sound shell (also known as Kulpi Watiku or Malaku Wilytja) is the most impressive as it is the best place to look at the cross-bedding that helps explain the depositional environment of the Mutitjulu Arkose. Unfortunately the interpretive signs say nothing about this, or the colour and composition of the rock. None of the signs are any good in explaining the geology. We ride on to the Kantju Gorge then backtrack to find the prescribed path around the north face.

The location of the base walk has changed significantly over time. Around the northern face of the Rock the trail used to follow the actual base of the Rock and visitors could marvel at the wonderful natural features on that side close up, exploring numerous caves in the process. The trail has been progressively moved further and further away. These days it follows the old circuit road and the edge of the old landing strip and you are sometimes over half a kilometre from the Rock. The views on the new sealed circuit road are comparable. Every 100m or so on the trail the view is marred by signs banning photographs of the Rock. Imagine the same practice being employed by other religions. It is utterly ridiculous and I take a few photos of the important geological features along here, in celebration of science and Western civilisation. There are so many lost opportunities for local culture to be showcased and celebrated along this section, instead it is closed off and censored, a metaphor for the closed minds of the Board and Parks Administrators, and their lack of imagination and faith in the goodwill of visitors. We ride around to the eastern end and then around the southern side. I stop here and there to evaluate alternate routes up. The valleys of Tjukiki Gorge, Metjan Rockhole and Mutitjulu all provide options perhaps slightly more challenging than the western spur but all achievable by experienced hikers. We drop the bikes off and head back to the resort for lunch.

In the afternoon the Climb beckons and we head back to the spur. I really want to check out the Last Logbook. Once again we climb up, the girls a bit more confident this time. The views are equally marvellous. At the summit monument the tupperware container is still there. I open it to find over 13 pages filled in. It is heartening to see this cultural practise restored, albeit for just a short time. I read through some of the comments of visitors from all the Australian states, Italy, Japan, Switzerland, England and Denmark:

- *Great Climb, against it closing.*
- *Awesome walk, keep it open for sure.*
- *Wonderful landscape, keep it open 100%.*
- *Highlight of the trip.*
- *Best way to experience Uluru.*
- *Goodbye Ayers Rock climb, it's been a wonderful experience.*
- *One of the most wonderful things I have done, this place is magic.*
- *I think this is what it would be like on Mars.*

The last one makes me smile. Matt Golombek would probably agree. We pack the book away and head back down. We have plans to visit Kings Canyon the next day so we leave the book to be collected on the day after, since the weather forecast is identical. The colour of the Rock on the way down is amazing – glowing hot red against the setting sun.

Figure 53: Dana, Zoe and myself at the summit monument. We'll be back in 20 years to do it again

On the way out to Kings Canyon we pass a long line of caravans, trailers and loaded 4WDs also heading east with an equal number of vehicles heading west. With the looming ban it seems many people are taking the opportunity to climb before the experience is taken away. We reach the canyon by late morning and after refuelling embark on the rim walk. The geology here is also quite remarkable. The day ends with a long drive back and a return just in time to watch the sunset over the Rock.

On our last day the intention is to retrieve the log book and visit the Olgas. The Climb is closed though, apparently it's too windy up there, although how you can tell without any instrumentation at the top is beyond me. It does not appear any windier than the previous day when the Climb was open. We leave the logbook to fate. Hopefully someone will follow the instructions and put it in the mail. We head west in the hire car.

Looking over the old guide maps, tourists used to be able to drive to the "Kata Tjuta Lookout". This is within the 36 domes. Charles Mountford described it as having the best views in Australia. You can no longer access it. Instead there is a viewing area along the road about 5.5km from the Olgas that is somewhat less impressive. Once again the interpretive signs let you down if you are interested in any hard science. We continue on, park at Walpa Gorge and start up the trail of this wonderful place, lamenting the ridiculous sanctions that prevent climbing any of the 36 domes. The southernmost would make an easy hike and provide extraordinary views back over the domes to Ayers Rock. If you search online you can find some photos taken by braver visitors less worried about the ridiculous censorship of the natural world. The adjacent Mt Olga should also be open for experienced climbers. The lost opportunities once again stand out. Perhaps the locals should take a trip to Italy and participate in some of the *Via Ferrata* walks. On their return they might think again about the possibilities at their Uluru. Time quickly runs out and we head back to the airport in time for our flight home.

I'll be back in another 20 years to climb again and to see if the enlightenment has finally made it to central Australia. It arrived briefly in the 1970s and seemed to be growing under the gentle guidance of Derek Roff, but was stalled and driven out by a chilling regression that followed the transfer of ownership and Federal takeover in 1985. With the advent of the internet and opportunities for the current Custodians to see and experience more of the world around them I am confident that in 20 years' time when I return I'll be welcomed with a climb is "open" sign, and several new *Via Ferrata* routes on the Rock and the Olgas. Who knows I might even find a copy of this book, and *Bromley Climbs Uluru* in the bookshop at the Cultural Centre?

The Last Logbook returned

A little over a month after returning from Ayers Rock I wrote to Steven Baldwin, one of the Managers at the Park, to let him know the release date for this book. We had been trading emails over the Climb issue for the past eight months or so and met briefly for coffee at the Cultural Centre café to discuss the looming ban when we visited in July. In the email Steve let me know a cleaning crew had found the logbook box about 20 metres from the summit monument with the envelope 50 metres away. Good Samaritan that he is, he arranged for the logbook to be returned.

About two weeks later the envelope arrived… a little worse for wear. The 192 page exercise book was full to the brim with signatures and positive statements about the climb and lamentations about the looming ban. The first names were ours, written at 4pm on 17 July. The last date was 21 August. Over that time more than 2,400 people had signed the book from 33 countries. The majority of comments were by families like ours, visiting the summit to gain an experience of the natural world and create memories to last a lifetime. The oldest climber was an 83-year-old woman, the youngest not yet born.

The logbook was scanned and sent to the Park Board and

Management. Perhaps if they read through the many names, the many families and nationalities, they may realise the full extent of the damage they are doing.

We leave the last word to the Herberts:

We made it! Came up 28/7/2018. This is just amazing! So glad we came with our kids, so disappointed they won't be able to climb with their kids.

ENDNOTES

[1] http://www.ntlis.nt.gov.au/placenames/view.jsp?id=10532

[2] Jeff Carter, *A Guide to Central Australia*, 1972.

[3] https://aiatsis.gov.au/sites/default/files/catalogue_resources/sutherland.s01.cs_.pdf

[4] 70,000 fewer visitors per annum spending $1000 each.

[5] https://www.australianarchaeologicalassociation.com.au/journal/new-radio-carbon-dates-for-kulpi-mara-rockshelter-central-australia/http://nature.com/articles/doi:10.1038/nature20125 BP- before present

[6] https://www.researchgate.net/publication/279298361_New_Radiocarbon_Dates_for_Kulpi_Mara_Rockshelter_Central_Australia

[7] C.P. Mountford, 1965. *Ayers Rock: Its People, Their Beliefs and Their Art*, Angus and Robertson.

[8] https://www.environment.gov.au/system/files/resources/4039eb6a-b3e7-49f0-b28d-200793b53057/files/uktnp-a4factsheet-pleasedontclimb-small.pdf

[9] Mountford, op. cit.

[10] https://australianmuseum.net.au/dingo

[11] G. Griffin, 2002. *Welcome to Aboriginal Land in Conservation and Mobile Indigenous Peoples: Displacement, Forced Settlement and Sustainable Development.* Dawn Chatty, Marcus Colchester (eds), Berghahn Books.

[12] https://theconversation.com/when-did-aboriginal-people-first-arrive-in-australia-100830

[13] W. C. Gosse's Explorations, 1873 http://gutenberg.net.au/ebooks13/1306451h.html

[14] https://trove.nla.gov.au/newspaper/article/164102648

[15] http://guides.naa.gov.au/records-about-northern-territory/part2/chapter16/16.8.aspx

[16] Journal of the Central Australian Exploring Expedition, 1889, under command of W.H. Tietkens, 1889. https://www.biodiversitylibrary.org/item/123122#page/3/mode/1up

[17] Background photo source Petticoat Safari 1957 courtesy Edna Bradley http://www.abc.net.au/radionational/programs/archived/hindsight/red-dust-travellers/3187412

[18] Baldwin Spencer Diary. http://spencerandgillen.net/objects/50ce72f4023fd7358c8a938f

[19] Album compiled by W H Tietkens mainly of his 1889 expedition in Central Australia. State Library NSW.

[20] Fig 44 Spencer and Gillian, 1912, *Across Australia*, p. 111, https://archive.org/stream/acrossaustralia01spenuoft?ref=ol#page/110/mode/2up

143

21 Ibid., p. 114.

22 Fig 47 Spencer and Gillian, op. cit., p. 123.

23 S. Breeden, 1994. *Uluru: Looking after Uluru-Kata Tjuta The Anagu Way*, p. 135.

24 https://trove.nla.gov.au/newspaper/article/16672742

25 https://trove.nla.gov.au/newspaper/article/45660118

26 This is the only mention of Bob Coulthard climbing. Perhaps McKenzie means Oliver?

27 https://collections.slsa.sa.gov.au/resource/B+38778

28 https://trove.nla.gov.au/newspaper/article/161789805

29 Ibid.

30 https://trove.nla.gov.au/newspaper/article/163041038

31 http://www.nma.gov.au/__data/assets/pdf_file/0005/176873 /Full_extract_1903_expedition.pdf

32 https://babel.hathitrust.org/cgi/pt?id=nyp.33433089899151;view=1up;seq=9

33 https://trove.nla.gov.au/newspaper/article/164102648

34 https://trove.nla.gov.au/newspaper/article/89655235

35 http://adb.anu.edu.au/biography/mackay-donald-george-7377

36 http://www.nma.gov.au/__data/assets/pdf_file/0004/176881 /Full_extract_1926_expedition.pdf

37 http://adb.anu.edu.au/biography/basedow-herbert-5151

38 https://trove.nla.gov.au/newspaper/article/143675557

39 http://collectionsearch.nma.gov.au/object/13565

40 https://trove.nla.gov.au/newspaper/article/164863006

41 https://trove.nla.gov.au/work/21721635

42 http://archives.samuseum.sa.gov.au/aa42/provlist.htm

43 *The Queenslander* (Brisbane, Qld: 1866-1939), p. 24. http://nla.gov.au/nla.news-page2368137

44 http://www.australiangeographic.com.au/topics/history-culture/2010/02/aviation-the-adventures-of-love-bird-and-diamond-bird/

45 https://trove.nla.gov.au/newspaper/article/16687592

46 *Daily Mercury*, 24/10/1930. Camera recovered from central Australia, p. 10

47 https://nla.gov.au/nla.obj-144673129

48 http://nla.gov.au/nla.obj-149249857/view

49 T. Healy (1995), *The Early Ascents of Mount Olga*, Burwood, self-published.

50 See reported list of climbers made by the Mt Olga-Ayers Rock research expedition of 1950, Arthur Groom in 1947 and Knox Grammar School in 1950.

51 http://www.wilmap.com.au/nt/atts/stuart_gaol.html

52 https://trove.nla.gov.au/newspaper/article/125174870

[53] Photo in Healy, op. cit.

[54] https://trove.nla.gov.au/newspaper/article/125174870

[55] https://trove.nla.gov.au/newspaper/article/65170334

[56] https://trove.nla.gov.au/newspaper/article/90950760

[57] https://trove.nla.gov.au/newspaper/article/37266096

[58] Robert Layton, 2001, *Uluru: An Aboriginal History of Ayers Rock*, p. 72.

[59] https://trove.nla.gov.au/newspaper/article/129296011

[60] https://trove.nla.gov.au/newspaper/article/2398118

[61] Photo in Healy, op. cit.

[62] https://web.archive.org/web/20151002082822/http://freepages.genealogy.rootsweb.ancestry.com/~robtan/AyersRockLogCli.html

[63] Correct date should be 7 March 1932.

[64] Knox Expedition to Ayers Rock Part 2, *Australasian Photo Review*, May 1951 p. 300.

[65] C.P. Mountford, 1965, *Ayers Rock Its People, Their Beliefs and Their Art*.

[66] https://trove.nla.gov.au/newspaper/article/244811236

[67] C.P. Mountford, 1949, *Brown Men and Red Sand*. Note Mountford's odd italisation of the word "all" adjacent the Foy party and reference to those that had not climbed. "I added my name to that illustrious throng," says Mountford. "On the highest point of the Rock some previous visitor had built a low cairn of stones around a glass screw-topped bottle, in which were the names of people, who in recent years had climbed Ayers Rock, and I'm afraid, many who had not."

[68] G.R.W. Latham, 1951, Knox Expedition to Ayers Rock, *Australasian Photo Review*, Vol. 58, No. 5 (1 May 1951). http://nla.gov.au/nla.obj-469507733/view?partId=nla.obj-469618908: "A short walk and a final climb completed the ascent to the cairn at the summit. Here, beneath a heap of stone, was a coffee jar serving to keep safe the names of previous climbers – practical and theoretical; some lists were obviously written by the one hand, and it is therefore open to doubt whether all those so-listed really climbed to the top."

[69] If readers are aware of one, we would love to see it!

[70] https://www.nfsa.gov.au/collection/curated/phantom-gold-rupert-kathner

[71] https://trove.nla.gov.au/newspaper/article/52266230

[72] https://trove.nla.gov.au/newspaper/article/47905717

[73] Morley Cutlack, Aubrey Hermann and Ludwig Louis Nudl, National Archives of Australia bar code: 777752.

[74] Cutlack Expedition 1. Application for entry into Aboriginal Reserve 2. General File National Archives of Australia 1936-1951 NAA: A431, 1951/197 Item bar code: 66560.

[75] https://trove.nla.gov.au/newspaper/article/234600061

[76] https://recordsearch.naa.gov.au/SearchNRetrieve/Interface/ViewImage.aspx?B=45801

[77] http://www.slsa.sa.gov.au/site/page.cfm?u=999

[78] http://www.amw.org.au/sites/default/files/memory_of_the_world/indige-nous-collections/mountford-sheard-collection.html

[79] An Australian youth among desert Aborigines: journal of an expedition among the Aborigines of central Australia by Lauri E. Sheard; with an introd. by Charles P. Mountford, https://catalogue.nla.gov.au/Record/709240

[80] https://trove.nla.gov.au/newspaper/article/130850602

[81] From Borgelt's Diary courtesy Lutheran Archives.

[82] Lutheran Archives.

[83] http://adb.anu.edu.au/biography/tjalkalyirri-tiger-15659

[84] Transcript of interview with AD (Derek) Roff MBE.

[85] Uluru–Kata Tjuta National Park Knowledge Handbook, Park-Aku Nintiring-anyi, June 2012, p. 109.

[86] https://trove.nla.gov.au/work/25774142

[87] https://trove.nla.gov.au/newspaper/article/18173504

[88] https://trove.nla.gov.au/newspaper/article/118722623

[89] https://trove.nla.gov.au/newspaper/article/48207266/4827154

[90] Knox Expedition to Ayers Rock, *Australasian Photo Review*, April 1951, Part 2 – May 1951, Part 3 – April 1951.

[91] See for instance *Sydney Morning Herald*, 13 September 1950, https://trove.nla.gov.au/newspaper/article/18173504

[92] *Red Horizon*, https://vimeo.com/55226156

[93] Ron Dingwell who assisted Len Tuit and drove the Blitz wagon has also provided a voice over to a similar video which provides a slightly different perspective on the expedition, available via Youtube. https://www.youtube.com/watch?v=ihp67gJZngs

[94] Beryl Miles, *The Stars My Blanket*, https://trove.nla.gov.au/work/19014568

[95] https://trove.nla.gov.au/newspaper/article/65168354

[96] https://catalogue.nla.gov.au/Record/3256848

[97] https://trove.nla.gov.au/newspaper/article/48207266/4827154

[98] Edna Bradley, *A Rock to Remember: a Memoir to Early Tourism to Uluru.*

[99] Uluru Kata-Tjuta National Park Plan of Management 2001.

[100] Bill Harney, *To Ayers Rock and Beyond*, Seal Books, 1974 edition.

[101] Email Mike Misso 5 April 2018.

[102] Ayers Rock-Mt Olga National Park Information Booklet, Northern Territory Reserves Board, 1967.

[103] For historical information and photos of the 1958 National Mapping field season across central Australia see https://www.xnatmap.org/adnm/ops/prog/rafgeosvy/08.htm

[104] See Australian Archives file:Ayers Rock Geodetic station. Item bar code 8241845

[105] https://www.xnatmap.org/

[106] Uluru (Ayers Rock-Mt Olga) National Park Plan of Management 1982.

[107] Alan White's Story of constructing the Uluru Climb Chain, Parks Australia.

[108] See Ayers Rock-Mt Olga park guide in Australian Archives file: Ayers Rock Geodetic station. Item bar code 8241845.

[109] C. Mountford, A. Roberts, *Your Guide to Ayers Rock*, Katajuta Publishers, 1961.

[110] See Australian Archives file:Ayers Rock Geodetic station. Item bar code 8241845.

[111] 1986 management plan.

[112] See Australian Archives file: Ayers Rock Geodetic station. Item bar code 8241845.

[113] *Ayers Rock & the Olgas*, text and photography by Derek Roff, 1979.

[114] Ibid.

[115] https://trove.nla.gov.au/work/31192405

[116] http://www.alicespringsnews.com.au/2015/01/27/the-rock-to-climb-or-not-to-climb/

[117] http://search.nt.gov.au/Coveo/Hubs.aspx?sh=Oral+History+Search Search on Derek Roff for summary.

[118] See https://en.wikipedia.org/wiki/File:Tjamiwa%27s_map_of_Uluru.jpg

[119] http://parlinfo.aph.gov.au/parlInfo/search/display/display.w3p;query=Id%3A%22chamber%2Fhansardr%2F1983-11-15%2F0020%22

[120] Australian National Parks Wildlife Service (1991). Uluru (Ayers Rock-Mt Olga) National Park: Visitors' guide/[Australian National Parks and Wildlife Service]. Canberra]: Australian National Parks and Wildlife Service.

[121] http://www.abc.net.au/news/2009-07-10/rudd-supports-uluru-climbers/1348742

[122] https://theconversation.com/why-we-are-banning-tourists-from-climbing-uluru-86755

[123] http://155.187.2.69/parks/uluru/culture/culture/index.html

[124] https://commons.wikimedia.org/wiki/File:Warning_sign_at_Uluru.JPG

[125] D.N. Young, N. Duncan, A. Camacho, P.A. Ferenczi and T.L.A. Madigan, 2002. *Ayers Rock, Northern Territory* (Second Edition). 1:250 000 geological map series explanatory notes, SG 52-8. Northern Territory Geological Survey, Darwin.

[126] I.P. Sweet, A.J. Stewart I.H. and Crick, *Uluru and Kata Tjuta: A Geological Guide*. Geoscience Australia, 2012.

[127] Steven Baldwin, Parks Australia. Currently regarded as the "Men's Cave" but reported by Mountford 1974 as the wet weather camp of the marsupial Mole Men.

[128] *Uluru and Kata Tjuta: A Geological Guide*, Geoscience Australia, 2012, p. 36.

[129] R.B. Thompson, *A Guide to the Geology and Landforms of Central Australia.*

[130] C.P. Mountford, *Ayers Rock: Its People, Their Beliefs and Their Art*, 1965.

[131] http://www.valeriebarrow.com/?p=54

[132] Arthur Groom, *I Saw a Strange Land: Journeys in Central Australia.*

[133] Ibid.

[134] http://www.stepform.com.au/as1675-2018/stairs-detail.html

[135] Uluru Park Guide, May 1981.

[136] See Starva, https://www.strava.com/segments/5852224

[137] Cyril E. Goode, Account of a dash to Ayer's Rock, Roma: *Accademia Internazionale Leonardo da Vinci*, 1972

[138] "Notes from the Rock", *Sydney Morning Herald*, 11 July 1970.

[139] Australia. Ayers Rock: The Kangaroo's Tail – An account of the first ascent in May 1973 by Keith Lockwood and Andrew Thomson, *Mountain Magazine*, February 1974.

[140] Uluru Guide, 1981.

[141] Ibid.

[142] Beryl Miles, *The Stars My Blanket*, 1954.

[143] 2010 Management Plan.

[144] 2012 Knowledge Handbook.

[145] Do not climb fact sheet.

[146] Various media sources quoting Parks Australia e.g., http://www.abc.net.au/news/2017-11-01/uluru-climbs-banned-after-unanimous-board-decision/9103512

[147] Introduction to Anangu culture for schools.

[148] http://www.abc.net.au/news/2018-07-04/japanese-tourist-dies-climbing-uluru/9937848

[149] https://www.talkinglifestyle.com.au/podcast/why-the-no-climb-decision-will-make-visiting-uluru-better/

[150] "Ayers Rock Death", *Sydney Morning Herald*, Monday, 28 May 1962.

[151] "Death Plunge on Rock: Sun blamed", *Sydney Morning Herald*, 24 December 1963.

[152] Various dates are mentioned in official records, 1964, 1965 and 1966.

[153] "Killed in Fall", *Sydney Morning Herald*, 19 May 1978.

[154] "Discovery of jacket vindicated Lindy", *The Australian*, 14 August 2010.

[155] Evil Angels, http://www.johnbryson.net/~/c/d/azaria-papers/AzBrett86.pdf

[156] Michael Chamberlain died wanting an apology and a plaque for baby Azaria News.com.au, 17 January 2017.

[157] "Climb time plan after Rock death", Canberra Times, 13 February 1991, https://trove.nla.gov.au/newspaper/article/129096042

[158] "Dead in the Heart", *The Bulletin*, 16 November 1999.

[159] http://www.abc.net.au/news/2010-04-25/climb-closed-after-mans-death-on-uluru/410934

[160] Uluru climb, health and safety review, May 2007. Report to Director National Parks.

[161] "Uluru death fuels call to close Rock climb", *NT News*, 27 April 2010.

[162] A 76-year-old Japanese tourist has died while climbing Uluru. https://www.sbs.com.au/news/japanese-tourist-dies-climbing-uluru

[163] "Japanese tourist dies in Uluru Park", *Sydney Morning Herald*, 7 January 2008.

[164] https://theconversation.com/whats-most-likely-to-kill-you-measuring-how-deadly-our-daily-activities-are-72505

[165] Ibid.

[166] https://en.wikipedia.org/wiki/Micromort

[167] https://theconversation.com/closing-uluru-to-climbers-is-better-for-tourism-in-the-long-run-86831

[168] 70,000 visitors fewer per annum, allowing $1000 per visitor spent locally at Yulara and Ayers Rock.

[169] http://www.smh.com.au/federal-politics/political-news/turnbull-government-decides-against-banning-tourists-from-climbing-on-uluru-20160411-go3qso.html

[170] https://www.nationalparks.nsw.gov.au/things-to-do/walking-tracks/wollum-bin-mount-warning-summit-track

[171] 4 x 4 Australia 01/06/2010 issue 317.